MASTER OF FIREPOWER
CHRONICLES OF ICONIC FIREARMS DESIGNERS

John Crump

For my wife who puts up with my crap. For Matthew and Nicky, who are the best sons I could ever ask for.

Table of Contents

Introduction 1

Henry Deringer 7

Johann Nicolaus von Dreyse 31

Eliphalet Remington 57

Oliver Winchester 83

Christian Sharps 109

Samuel Colt 135

Louis-Nicolas Flobert 161

Benjamin Tyler Henry 187

Wilhelm Mauser 213

Hiram Maxim 239

Georg Luger 265

John Moses Browning 291

John Thompson 317

Hugo Schmeisser 343

John Garand 369

Aimo Lahti 395

Yisrael Galili 419

Mikhail Kalashnikov 445

Eugene Stoner 469

Gaston Glock 495

Epilogue 521

Introduction

Unveiling The Legacy of Historical Firearms Makers

The world of firearms has been shaped by the innovative minds and skilled craftsmanship of legendary weaponsmiths throughout history. These pioneering individuals have left an indelible mark on the evolution of firearms, revolutionizing design and manufacturing techniques that continue to influence modern weaponry. One such pioneer is Samuel Colt, an American inventor who revolutionized firearm production with his creation of the Colt revolver in the early 19th century.

His groundbreaking design, incorporating a revolving cylinder that held multiple rounds, provided unparalleled firepower and reliability. This innovation not only transformed personal defense but also had a profound impact on military strategies. Another iconic figure in firearm history is John Moses Browning, an American gun designer whose inventions include the famous Browning Automatic Rifle (BAR) and M1911 pistol. Browning's meticulous attention to detail and relentless pursuit of perfection resulted in highly reliable firearms that were widely adopted by military forces around the world.

Across the Atlantic, Louis-Nicolas Flobert made significant contributions to firearm development with his creation of rimfire ammunition in 1845. His invention paved the way for smaller-caliber firearms suitable for both sport shooting and self-defense purposes.

***

Throughout history, firearms have played a pivotal role in shaping societies, wars, and even revolutions. Behind these iconic weapons are the master craftsmen who dedicated their lives to perfecting their craft and leaving behind a lasting legacy. Exploring the impact of historical firearms makers sheds light on their influence not only on weaponry but also on artistry, innovation, and technological advancements.

These remarkable artisans were more than just gunsmiths; they were pioneers who pushed boundaries and revolutionized firearm design. Their expertise went beyond mere functionality, with intricate engravings adorning the metalwork, transforming these weapons into true works of art. The craftsmanship displayed by these makers was unparalleled and has become a benchmark for excellence in gun making. Moreover, historical firearms makers' innovations paved the way for significant developments in military tactics and warfare.

From flintlock to breech-loading mechanisms, these inventors constantly sought ways to improve accuracy, reloading speed, and overall performance. Their contributions had a direct impact on shaping battles throughout history. Beyond warfare applications, historical firearms makers also left an indelible mark on popular culture. Iconic guns like Colt revolvers or Winchester rifles have become symbols of American frontier mythology or cowboy legends.

***

Throughout history, certain gunmakers have risen above the rest, leaving an indelible mark on the world of firearms. These legendary artisans and craftsmen have not only shaped the evolution of weaponry but have also become synonymous with excellence in design, precision engineering, and innovation. One such legendary gunmaker is Samuel Colt, whose name became synonymous with revolvers. His groundbreaking invention, the Colt Revolver, revolutionized firearms manufacturing and played a pivotal role in shaping America's Wild West.

Colt's commitment to quality and his relentless pursuit of perfection made his firearms highly sought after by lawmen, outlaws, and soldiers alike. John Moses Browning is another iconic figure in the realm of gun making. With over 128 firearm patents to his name, Browning's designs became renowned for their reliability and effectiveness. His inventions spanned various platforms including rifles, shotguns, and pistols.

The timeless designs he created continue to influence modern-day firearms. In Europe, Johann Nicolaus von Dreyse stands as a legend for his creation of the needle gun - a significant advancement in military rifle technology during the mid-19th century. This firearm was adopted by several European armies due to its superior range and rapid-fire capability.

***

The history of firearms is a testament to human ingenuity and technological advancement. From the early days of flintlock mechanisms to the cutting-edge innovations of modern firearms, this evolution has shaped not only the weapons themselves but also the course of history. The journey begins with flintlock firearms, which dominated battlefields for centuries. These muzzle-loading guns utilized a flint striking against steel to ignite gunpowder and propel a lead ball toward its target.

Despite their simplicity, flintlocks were reliable and effective weapons that shaped warfare during the 17th and 18th centuries. As time progressed, so did firearm technology. Percussion caps replaced flintlocks in the early 19th century, allowing for faster ignition and increased reliability. This development paved the way for an array of legendary firearms makers who sought to push boundaries further. Innovators like Samuel Colt revolutionized firearm design with his patented revolving cylinder mechanism, enabling rapid-fire capabilities in handguns.

Meanwhile, John Moses Browning's genius gave birth to iconic semi-automatic pistols and rifles that set new standards in terms of reliability and firepower. Today, we find ourselves amidst a golden age of firearms manufacturing.

***

Historical firearms hold a significant place in our cultural heritage, embodying the craftsmanship and innovation of past generations. These masterpieces not only showcase the artistry and engineering prowess of their makers but also provide invaluable insights into our history. Preserving the legacy of these firearms makers is essential in ensuring that their contributions to our society are recognized and appreciated. By celebrating the enduring legacy of historical firearms makers, we acknowledge their role in shaping the development of weaponry, as well as their impact on military strategies and societal norms.

The meticulous attention to detail, precision manufacturing techniques, and artistic embellishments displayed by these craftsmen have left an indelible mark on firearm design. Moreover, understanding the historical context surrounding these firearms allows us to appreciate how they have shaped our society today. From revolutionary advancements in firearm technology to weapons that played a pivotal role in significant historical events, each piece carries its own unique story waiting to be unveiled.

Preserving this rich heritage involves not only showcasing these firearms but also documenting and studying them extensively. Museums, private collectors, and organizations dedicated to firearm history play a vital role in safeguarding this legacy for future generations.

Henry Deringer
1786-1868

In the world of firearms, there are some designs that stand out as true game-changers. One such iconic weapon is the Deringer pistol, named after its creator, Henry Deringer. Born in 1786 in Easton, Pennsylvania, Deringer would go on to revolutionize the firearm industry with his innovative design and exceptional craftsmanship.

During the early 19th century, firearms were primarily large and cumbersome. They lacked portability and were often difficult to conceal. This posed a significant challenge for individuals who sought personal protection or needed a reliable weapon in emergency situations. Recognizing this need for a compact yet effective firearm, Henry Deringer set out to create something truly groundbreaking.

Deringer's first major breakthrough came in 1825 when he patented his pocket-sized pistol design. Unlike its predecessors, which were typically flintlock or percussion cap pistols with multiple barrels, Deringer's creation featured a single-shot mechanism with a concealed hammer. This made it much easier to handle and reduced the size significantly without compromising firepower.

What truly set the Deringer pistol apart from other firearms of its time was its size. Measuring only around six inches long with an overall weight of just under one pound, it could easily fit into a pocket or be discreetly carried by anyone requiring protection. The compact size made it an ideal choice for gamblers, travelers, law enforcement officers, and even women who wanted a reliable means of self-defense.

Deringer's revolutionary design did not sacrifice firepower either. The pistol was chambered for .41. caliber rounds—considerably larger than what was commonly used in similar-sized firearms at that time—ensuring adequate stopping power when needed most.

Another key feature that contributed to the success of this game-changing design was its simplicity and reliability. By eliminating unnecessary components like multiple barrels or complicated firing mechanisms, Deringer created a firearm that was easier to manufacture and maintain. This made it more affordable and accessible to a wider range of individuals, cementing its popularity.

The impact of Henry Deringer's innovative design was felt far beyond the borders of the United States. The pistol became highly sought after both domestically and internationally, with numerous imitations flooding the market. However, none could truly match the quality and craftsmanship of an original Deringer pistol.

Henry Deringer's revolutionary design forever changed the landscape of firearms. His pocket-sized pistol provided individuals with an effective means of personal protection in a compact and easily concealable package. The simplicity, reliability, and power packed into this game-changing weapon ensured its place in history as one of the most iconic firearms ever created. Even today, more than two centuries later, the Deringer pistol continues to captivate collectors and enthusiasts alike with its timeless design and historical significance.

***

In the world of firearms, certain designs and innovations have revolutionized the industry, forever leaving their mark on history. One such iconic weapon is the Deringer pistol, a small caliber firearm created by Henry Deringer in the mid-19th century. The introduction of this compact and concealable handgun marked a new era in firearms technology, bringing about significant changes in personal defense and law enforcement practices.

Before Deringer's invention, handguns were typically large and cumbersome, making them less practical for everyday carry. However, with his innovative design, Deringer managed to create a lightweight pistol that could be easily concealed within a coat pocket or hidden beneath clothing. This breakthrough allowed individuals to possess a weapon discreetly and increased their ability to defend themselves effectively.

One of the key features that set Deringer's pistols apart was their small caliber ammunition. While earlier handguns relied on larger calibers for increased stopping power, Deringer understood that smaller projectiles could still be lethal at close range while offering improved accuracy. By utilizing smaller calibers such as .41 or .44, his pistols provided users with greater control over their shots while maintaining sufficient firepower.

The impact of these small caliber weapons was felt not only by civilians seeking personal protection but also by law enforcement agencies across the United States. The compact size and ease of concealment made them an attractive choice for officers who needed to carry a firearm discreetly during undercover operations or when off duty. In this way, Deringer's creation became an essential tool for agents tasked with maintaining public safety.

Furthermore, due to its popularity among law enforcement personnel and civilians alike, the Deringer pistol became synonymous with personal defense during this period. Its reputation as a reliable self-defense tool grew rapidly as stories circulated about individuals successfully defending themselves against threats using these compact firearms. The Deringer pistol became an icon of personal empowerment, empowering individuals to take control of their own safety.

Beyond its immediate impact on personal defense, the Deringer pistol also influenced future firearms designs. Its compact size and lightweight construction served as a blueprint for subsequent concealed-carry weapons, inspiring manufacturers to develop even more compact and efficient handguns. This led to the development of numerous small caliber pistols that are still popular today.

Henry Deringer's creation of the small caliber Deringer pistol brought about a new era in firearms technology. By introducing a compact and concealable handgun that utilized smaller calibers, he revolutionized personal defense and law enforcement practices. The impact of this innovation was far-reaching, empowering individuals to protect themselves discreetly while inspiring further advancements in firearms design. The legacy of the Deringer pistol lives on, forever changing the game in the world of firearms.

***

In the early 19th century, firearms were predominantly large, cumbersome weapons that lacked practicality when it came to personal self-defense. However, all that changed with the introduction of the iconic Deringer pistol by Henry Deringer. This revolutionary pocket-sized handgun brought forth a new era of concealable power, forever altering the way people perceived and utilized firearms.

The Deringer pistol was renowned for its compact design and lightweight nature, making it an ideal choice for those seeking a discreet and easily concealable weapon. Measuring only five to six inches in length, these pistols could effortlessly fit into a pocket or be concealed within a hidden compartment. Its small size allowed individuals to carry it inconspicuously, granting them an advantage in surprise attacks or situations where the open display of arms would be inappropriate.

Henry Deringer's innovation not only lay in creating a compact firearm but also in its exceptional power, considering its size. The Deringer pistol boasted impressive stopping power and accuracy despite being significantly smaller than traditional firearms of the time. This remarkable feat was achieved by employing rifling techniques on its barrels, which enhanced bullet stability and accuracy while increasing muzzle velocity.

The result was a deadly weapon that could deliver lethal force at close range with precision.

The emergence of the Deringer pistol as a popular choice for personal protection can be attributed to several factors beyond its compactness and firepower. During this period, society was experiencing an increase in urbanization and industrialization, leading to crowded cities with rising crime rates. People sought reliable means to defend themselves against potential threats lurking in dark alleys or during their daily commutes.

Moreover, cultural shifts towards individualism and self-reliance during this era also contributed to the popularity of concealed firearms like the Deringer pistol. People valued their independence and wanted control over their personal safety. The ability to carry a concealed weapon provided them with a sense of security and empowerment, allowing them to face the uncertainties of urban life with confidence.

The Deringer pistol's influence extended beyond personal protection. Its compact design made it an attractive choice for gamblers, adventurers, and even women seeking a reliable firearm they could carry discreetly. Countless stories and legends emerged of individuals who relied on the concealable power of the Deringer pistol to defend themselves or ensure their survival in dire situations.

Henry Deringer's creation of the pocket-sized handgun revolutionized the firearms industry and forever changed the way people perceived self-defense weapons. The Deringer pistol offered concealable power that allowed individuals to carry a compact yet lethal firearm with ease. Its popularity stemmed not only from its discreet nature but also from its exceptional accuracy and stopping power for its size.

In an era marked by urbanization, rising crime rates, and cultural shifts towards individualism, the Deringer pistol provided people with a means to protect themselves while retaining a sense of control over their personal safety.

***

When it comes to compact firearms, few have made as significant an impact as the iconic Deringer pistol. Invented by Henry Deringer in the mid-19th century, this single-shot wonder revolutionized the world of handguns with its sleek design and remarkable efficiency. Let us delve into the story behind this groundbreaking firearm and explore how it forever changed the game.

Henry Deringer was a renowned American gunsmith who had already gained recognition for his high-quality pocket pistols before he introduced his most influential creation, the Deringer pistol. What set this firearm apart from its contemporaries was its compactness without compromising power or accuracy. Measuring only around five to six inches in length, it became a favorite among those seeking a discreet yet effective self-defense weapon.

The single-shot mechanism of the Deringer pistol was another hallmark that redefined compact firearms. Unlike other handguns of that time, which often featured multiple barrels or revolving chambers, Deringer's design simplified things significantly by employing just one barrel and one round per load. This simplicity not only reduced manufacturing costs but also increased reliability and ease of use.

The craftsmanship displayed in each Deringer pistol was exceptional, reflecting Henry Deringer's dedication to producing high-quality firearms. He meticulously handcrafted each piece using top-grade materials such as brass or silver for frames and walnut wood for handles. The attention to detail extended to every aspect, including engraving and decorative elements on the surface of these firearms.

While initially intended for self-defense purposes, the popularity of Deringer pistols quickly spread beyond personal protection due to their reliability and aesthetic appeal. Their small size made them ideal companions for gamblers, lawmen, adventurers, and even women seeking concealed carry options during an era when societal norms restricted their choices.

One notable aspect that contributed to the fame of these guns was their association with historical figures. The Deringer pistol gained iconic status when it became forever linked to the assassination of President Abraham Lincoln. John Wilkes Booth, the infamous assassin, used a Deringer pistol to carry out his heinous act, forever cementing its place in history.

As time went on, imitations and adaptations of the original Deringer pistol emerged, further solidifying its influence on compact firearms. Many manufacturers sought to replicate its success by creating similar single-shot pocket pistols that echoed Deringer's design principles.

Henry Deringer's creation of the iconic single-shot Deringer pistol reshaped the world of compact firearms. Its sleek design, single-shot mechanism, and exceptional craftsmanship set a new standard for efficiency and reliability in handguns. Beyond its practicality as a self-defense weapon, this firearm became an embodiment of style and sophistication. Whether associated with historical events or appreciated for its ingenuity alone, the impact of the Deringer pistol remains undeniable even today—a testament to Henry Deringer's game-changing contribution to firearms history.

***

The Deringer pistol, with its compact size and unmistakable design, has become an iconic symbol of American firearms history. This small, single-shot handgun was first introduced by Henry Deringer in the mid-19th century and quickly gained popularity among civilians and law enforcement officers alike. Today, the Deringer pistol remains a testament to Henry Deringer's innovation and influence on the firearms industry.

Henry Deringer was a renowned gunsmith based in Philadelphia during the early 1800s. Born in 1786, he honed his skills under the guidance of his father, a well-known gunmaker. In 1825, Henry patented his first pocket-sized pistol design, which featured a percussion lock mechanism. This innovative firearm would later be known as the "Deringer" or "Derringer" pistol.

The original Derringer pistol was chambered for .41 caliber ball ammunition and had a barrel length of around three inches. It weighed approximately one pound and featured a single-shot design that required manual reloading after each shot. What made this weapon stand out from others of its time was its compactness - it could easily be concealed in a coat pocket or hidden inside clothing.

The success of the original Derringer pistol led to numerous imitations by other gun manufacturers. However, due to their immense popularity, these copies were often referred to as "derringers," regardless of their actual origin or manufacturer. The term "derringer" eventually became synonymous with any small-sized pocket pistol.

Over time, Henry Deringer continued to refine his designs, introducing various improvements to enhance both functionality and aesthetics. One notable development was the addition of an automatic extractor mechanism that simplified cartridge removal after firing. Another improvement involved incorporating rifling into the barrels for increased accuracy.

Henry's son continued his father's legacy after his death in 1868, further expanding the Deringer pistol line. The Deringer family's firearms business eventually merged with the Colt Company in 1903, consolidating their expertise and contributing to the ongoing development of firearms technology.

The Deringer pistol played a significant role in American history during its prime. Its small size made it a popular choice for concealed carry and self-defense purposes. Notably, this pistol gained notoriety when John Wilkes Booth used a .44 caliber Deringer to assassinate President Abraham Lincoln in 1865.

Today, the Derringer pistol remains an iconic symbol of American firearms history. Although advancements in firearm technology have rendered it obsolete as a military weapon, its place in popular culture and historical significance cannot be overlooked. Numerous replicas and reproductions of this classic firearm are still produced today, catering to collectors and enthusiasts who appreciate its unique design and historical importance.

The Deringer pistol revolutionized small-sized handguns with its compactness and functionality. Henry Deringer's innovative designs continue to inspire gunmakers even today, cementing his place as one of the pioneers who changed the game of firearms manufacturing forever.

***

When it comes to firearms, size does not always dictate power. This notion is particularly exemplified by the iconic Deringer pistol, a compact and concealable weapon that revolutionized the way handguns were perceived during its time. Invented by Henry Deringer in the mid-19th century, this small yet deadly firearm packed a punch far beyond what its diminutive size would suggest. The first aspect to consider when unpacking the lethal potential of the Deringer pistol is its design and construction.

Measuring only around 6 inches in length and weighing just under a pound, this single-shot handgun was easy to handle and conceal, making it an ideal choice for personal defense or as a backup weapon. The compactness of the Deringer pistol allowed it to be easily hidden in pockets, sleeves, or even discreetly carried in handbags or holsters. What truly set the Deringer pistol apart from other firearms of its time was its caliber.

While larger pistols commonly used .36 or .44 caliber ammunition, Henry Deringer ingeniously designed his namesake weapon to fire smaller .41 or .45 caliber rounds. Despite their reduced size, these bullets were capable of inflicting significant damage due to their high velocity and energy transfer upon impact. The combination of a small gun firing relatively large bullets created a potent force that could incapacitate an assailant with just one shot.

Another crucial factor contributing to the lethal potential of the Deringer pistol was its accuracy at close range. Although it lacked sights for aiming as larger pistols did, this firearm's short barrel facilitated precise shooting within limited distances. By minimizing recoil and maintaining control over muzzle rise during firing, users could deliver accurate shots even without extensive marksmanship skills. Additionally, reloading speed played a vital role in determining the effectiveness of any firearm during that era, and the Deringer pistol excelled in this regard.

With a simple break-action design, the user could easily open the pistol, extract the spent cartridge, and load a fresh round into the chamber. Although it was still a single-shot weapon, this quick reloading process minimized downtime between shots, making it possible to fend off multiple attackers or provide cover fire in dire situations. While some may underestimate the lethal potential of such a small firearm, history has proven time and again that size is not always indicative of power.

The Deringer pistol's compactness and concealability allowed it to be carried inconspicuously while providing its wielder with an effective means of self-defense or offense. From its innovative caliber choice to its accuracy at close range and efficient reloading process, Henry Deringer's creation changed the game by showcasing how a small firearm could possess deadly potential beyond imagination. unpacking the lethal potential of the Deringer pistol reveals that this diminutive weapon was far from being just another small gun.

Its compact design, combined with its powerful caliber ammunition and accuracy at close range, made it a formidable firearm in personal defense situations.

***

Throughout history, firearms have played a significant role in shaping societies and cultures. One particular type of firearm that holds a unique place in history is the concealable pistol. These compact and easily hidden weapons, such as the iconic Deringer pistol invented by Henry Deringer, not only revolutionized personal protection but also had a profound impact on fashion, crime, and self-expression.

In the early 19th century, carrying a firearm was as much about making a fashion statement as it was about self-defense. Wealthy gentlemen often wore pistols prominently displayed in holsters attached to their belts or kept them readily accessible in their pockets. However, this changed with the introduction of the concealable pistol. These small handguns could be easily concealed within clothing or hidden in accessories like walking sticks or ladies' purses.

This shift from openly displaying weapons to hiding them signaled a new era where personal safety became more discreet and secretive.

The cultural significance of concealable pistols extended beyond mere functionality; they became symbols of power, status, and rebellion. In an era when social norms dictated that individuals settle disputes through duels or confrontations, carrying a concealed weapon represented readiness for any challenge that might arise. It conveyed an image of strength and fearlessness while simultaneously challenging traditional notions of honor and chivalry.

The popularity of concealable pistols also had its dark side – they quickly became tools for criminals and outlaws seeking an advantage during illegal activities. Their small size made them easy to hide while providing enough firepower to pose serious threats to law enforcement officers or innocent civilians alike. The infamous American outlaw Jesse James was known for his use of concealable pistols during his criminal exploits, adding an aura of danger and mystery around these firearms.

Beyond their impact on fashion trends and crime rates, concealable pistols held immense cultural significance as symbols of individuality and self-expression. Owning and carrying a concealed weapon became a statement of personal freedom, indicating a person's desire to protect themselves and their loved ones. It provided individuals with a sense of empowerment, enabling them to take control of their own safety in an unpredictable world.

The cultural significance of concealable pistols cannot be understated. From being fashion accessories that showcased wealth and status, they evolved into deadly weapons that altered the dynamics of crime and law enforcement. They represented power, rebellion, and personal freedom while challenging societal norms. Henry Deringer's invention revolutionized personal protection by transforming firearms into discreet tools that forever changed the game.

***

Throughout history, the Deringer pistol has gained notoriety for its compactness, concealability, and deadly effectiveness. With its unique design and powerful impact, this iconic firearm has been involved in various legendary cases and wielded by infamous users. Let us delve into some of the most notable moments in Derringer's pistol history.

One of the most renowned incidents involving a Derringer pistol occurred on April 14, 1865. John Wilkes Booth, an American actor, and Confederate sympathizer, assassinated President Abraham Lincoln at Ford's Theatre in Washington, D.C. Booth used a Philadelphia Deringer to shoot Lincoln in the back of the head while he was watching a play. This tragic event forever linked the Derringer pistol to one of America's darkest chapters.

Another infamous user of the Deringer pistol was Charles J. Guiteau. In 1881, Guiteau shot President James A. Garfield at a train station in Washington, D.C., using a British Bulldog variant of the Derringer pistol. The assassination attempt resulted in Garfield's death several months later due to infection from his wounds.

The Wild West was no stranger to encounters involving these small yet lethal firearms. One such case involved Wild Bill Hickok—an iconic figure known for his gunfighting skills—and his fateful encounter with Jack McCall on August 2, 1876. Hickok was playing poker when McCall entered the saloon and fired two shots at him from a Colt .45 revolver. Unfortunately for Hickok, neither shot hit their mark as he held two pairs—aces and eights—in what became known as "the dead man's hand."

In response to this failed assassination attempt by McCall, Hickok's friend Charlie Utter drew his own weapon—a .41 caliber Rimfire Derringer pistol—and shot McCall in the head, killing him instantly.

The Derringer pistol also played a significant role in the assassination of President William McKinley. On September 6, 1901, anarchist Leon Czolgosz attended a public reception in Buffalo, New York. Concealing a .32 caliber Iver Johnson revolver within a handkerchief, he approached McKinley and fired two shots. One bullet grazed the president's shoulder, but the other entered his abdomen. Despite receiving medical attention, McKinley succumbed to his injuries eight days later.

Beyond these notable cases involving infamous users, the Derringer pistol became popular among gamblers and women seeking personal protection during the late 19th century. Its small size allowed for discreet concealment within pockets or purses—a feature highly valued by those who needed an easily accessible means of self-defense.

The legendary cases and infamous users associated with the Derringer pistol have solidified its place in history as an iconic firearm that changed the game forever. From presidential assassinations to Wild West shootouts, this compact yet deadly weapon has left an indelible mark on our collective memory.

***

When it comes to iconic firearms in American history, the Deringer pistol undoubtedly holds a special place. Invented by Henry Deringer in the mid-19th century, this compact and concealable handgun revolutionized the world of firearms. While its immediate impact was profound, its influence continues to reverberate through time, shaping the design and functionality of modern-day handguns.

One of the most significant contributions made by Henry Deringer was the concept of a small, easily concealable firearm. Before his invention, handguns were often large and bulky, making them difficult to carry discreetly. The Deringer pistol changed this dynamic entirely. With its compact size and lightweight construction, it became a favorite among law enforcement officers and civilians alike who sought a reliable self-defense weapon that could be easily hidden from view.

Today, this concept remains at the forefront of firearm design. Manufacturers continue to produce small and concealable handguns that owe their existence directly to Deringer's pioneering work. Whether it's for personal protection or professional use by law enforcement agencies, compact pistols influenced by Deringer's invention have become an integral part of modern society.

Another enduring legacy left behind by Henry Deringer is his innovative approach to manufacturing techniques. In his time, he employed advanced production methods such as interchangeable parts and precision machining—a rarity for firearm manufacturers during that era. These techniques not only enhanced efficiency but also improved overall quality control.

Deringer's commitment to precision manufacturing has become a cornerstone of contemporary firearms production as well. Modern gun manufacturers utilize advanced technologies like computer numerical control (CNC) machining and automated assembly lines to ensure consistent quality across their product lines. The emphasis on precision has not only made firearms more reliable but has also allowed for greater customization options for individual users.

Furthermore, Henry Deringer's contribution extended beyond technical innovations. He was also responsible for popularizing the concept of personal protection, a concept that remains central to firearms ownership today. By creating a compact and reliable handgun, Deringer empowered individuals to take responsibility for their own safety—a notion that has become deeply ingrained in American culture.

The legacy of Henry Deringer and his invention is far from forgotten. His pioneering work in designing a small and concealable handgun revolutionized firearm manufacturing and continues to shape the industry today. The concept of compact handguns, precision manufacturing techniques, and the idea of personal protection all owe their existence to Deringer's groundbreaking contributions. As we look at modern firearms, it becomes evident that Henry Deringer's influence lives on—an enduring testament to his ingenuity and impact on the world of firearms.

***

Antique firearms have always captivated collectors, but few possess the allure and historical significance of the iconic Deringer pistol. As a pioneer in the world of small concealable handguns, Henry Deringer revolutionized firearms design and left a lasting impact on American history. For those fortunate enough to own one of these prized artifacts, understanding their value and preserving them becomes paramount.

In this collector's corner, we delve into the intricacies of valuing and preserving antique Deringer pistols.

Valuing an antique firearm requires a comprehensive analysis that encompasses various factors. The first is authenticity – ensuring that the pistol is indeed an original creation by Henry Deringer himself or his renowned successors. Examining the maker's marks, serial numbers, and other unique identifiers can help verify its authenticity.

Condition plays a crucial role in determining value as well. Collectors seek pistols with minimal damage or wear, such as intact grips, undamaged barrels without rust or pitting, and functional mechanisms. Original finishes are highly prized; however, refinishing can significantly diminish a pistol's worth unless executed professionally to match its original appearance.

Rare variants or special editions command higher prices within the collector market. For instance, early models with unusual features like silver inlays or engraving fetch premium values due to their scarcity and aesthetic appeal. Historical provenance adds another layer of value; if a particular pistol has connections to famous individuals or notable events from history, its worth increases exponentially.

Preservation is equally vital for collectors aiming to safeguard their antique Deringers for future generations. Implementing proper storage techniques helps prevent damage caused by moisture fluctuations or exposure to harmful elements like sunlight or corrosive substances. Display cases with regulated humidity levels protect against rust formation while keeping them safe from accidental handling.

Regular maintenance ensures that these delicate firearms remain operational while minimizing potential damage from aging mechanisms. Applying a light coating of preservation oil to all metal surfaces helps prevent rust, while periodic inspection and cleaning of internal components maintain their functionality. However, it is crucial to consult with experts before attempting any disassembly or cleaning to avoid unintentional damage.

When handling antique firearms, collectors should exercise great care and adhere to safety protocols. Wearing gloves while touching Deringer pistols prevents the transfer of oils from hands that may contribute to corrosion. Additionally, firearms should always be treated as loaded weapons and never pointed toward anyone or discharged without proper knowledge and training.

To preserve the historical significance of these iconic pistols, collectors are encouraged to document their firearms meticulously. Creating a comprehensive record that includes photographs, detailed descriptions, provenance information, and any historical documentation significantly enhances their value and provides invaluable insights for future generations.

valuing and preserving antique Deringer pistols require a combination of expertise, attention to detail, and an appreciation for their historical significance. By carefully assessing authenticity, condition, rarity, and provenance while implementing proper storage techniques and maintenance practices, collectors can ensure these remarkable artifacts endure for generations to come – continuing Henry Deringer's legacy as a game-changer in the world of firearms design.

Johann Nicolaus von Dreyse
1787-1867

In the annals of history, there are individuals who have left an indelible mark on society with their groundbreaking innovations and inventions. Among these remarkable figures is Johann Nicolaus von Dreyse, a German inventor renowned for revolutionizing firearms technology with his pioneering work on breech-loading rifles. Often referred to as the "Father of Breech-loading Rifles," Dreyse's contributions not only transformed the field of weaponry but also played a pivotal role in shaping military tactics and strategies.

Born on November 20, 1787, in Sömmerda, Germany, Johann Nicolaus von Dreyse grew up amidst the tumultuous era of Napoleonic wars that swept through Europe. Inspired by his father's passion for gunsmithing, young Johann developed a keen interest in firearms from an early age. His relentless curiosity and natural aptitude for mechanics soon set him on a path that would forever change the course of modern warfare.

Dreyse's breakthrough came in 1824 when he successfully designed and patented his first breech-loading rifle. Unlike traditional muzzle-loading firearms that required time-consuming reloading from the muzzle end after each shot, Dreyse's innovative rifle allowed for quick loading through the breech—a significant advancement that enhanced both efficiency and rate of fire. This revolutionary design not only improved accuracy but also significantly reduced reloading times during combat situations.

Following his initial success, Dreyse continued to refine his invention by introducing further improvements to enhance reliability and ease of use. One such improvement was his self-contained metallic cartridge—an ingenious combination of bullet, powder charge, and primer encased within a single brass or iron unit—which eliminated the need for separate loading components. This innovation further streamlined reloading processes while ensuring consistent performance across multiple shots.

The exceptional quality and effectiveness of Dreyse's rifles soon garnered international recognition. His designs were adopted by various military forces across Europe, including the Prussian Army, which became a staunch advocate of his firearms. During the mid-19th century, Dreyse's rifles played a pivotal role in several significant conflicts, notably the Austro-Prussian War and the Franco-Prussian War. The Prussian army's decisive victories in these conflicts can be partially attributed to the superior firepower and efficiency provided by Dreyse's innovative firearms.

Beyond his contributions to weaponry, Johann Nicolaus von Dreyse also demonstrated an unwavering commitment to social welfare and employee well-being. He established factories that provided fair wages and implemented progressive labor practices for his workers—an approach far ahead of its time. As we delve deeper into Johann Nicolaus von Dreyse's life and achievements, we will explore not only his technological breakthroughs but also uncover the man behind these innovations—a visionary inventor with a profound impact on both warfare and societal progress.

Through our journey into his remarkable life story, we will gain insights into how one individual can shape history through their ingenuity, perseverance, and dedication to advancing human capabilities. References:

***

Johann Nicolaus von Dreyse, renowned as the Father of Breech-loading Rifles, was born on November 20, 1787, in Sömmerda, a small town in Thuringia, Germany. His early life was marked by various experiences and influences that played a pivotal role in shaping his passion for firearms and ultimately revolutionizing the field of military weaponry. Growing up in a modest household, Dreyse was exposed to the craftsmanship of his father, Johann Christoph Dreyse, who worked as a locksmith.

The young Dreyse often spent hours observing his father at work, fascinated by the intricate mechanisms and precision required to create functional objects. This exposure to mechanical engineering at an early age planted the seeds of curiosity within him. Another influential aspect of Dreyse's background was his family's military tradition. His grandfather had served as a gunsmith for the Prussian army during the Seven Years' War.

Stories about his grandfather's involvement in crafting firearms for soldiers were frequently shared within the family circle. These tales ignited a deep fascination with weapons and their potential impact on warfare. Dreyse's formal education began at a local school where he demonstrated exceptional aptitude for mathematics and physics. His teachers noticed his keen interest in mechanics and encouraged him to pursue further studies in engineering.

Recognizing their son's potential, Dreyse's parents supported his aspirations and enrolled him at the prestigious Gewerbeinstitut (Institute of Technology) in Berlin. During his time at Gewerbeinstitut, Dreyse immersed himself in various subjects related to mechanical engineering. He studied under renowned professors who further nurtured his technical skills while broadening his knowledge base. It was during this period that he delved deeper into understanding firearms' mechanisms and explored ways to enhance their efficiency.

The Napoleonic Wars, which ravaged Europe during Dreyse's formative years, also played a significant role in shaping his passion for firearms. The sight of soldiers equipped with cumbersome and slow-loading muskets deeply troubled him. Witnessing the inefficiency of these weapons firsthand motivated Dreyse to explore alternative designs that could revolutionize the battlefield. Dreyse's experiences as a soldier himself further fueled his desire to improve firearms.

He served in the Prussian army, where he witnessed the devastating consequences of outdated weaponry in combat. These experiences solidified his resolve to develop a breech-loading rifle that would be faster and more reliable than the traditional muzzle-loaded muskets. Johann Nicolaus von Dreyse's early life was filled with influences that shaped his passion for firearms and laid the foundation for his groundbreaking innovations in weapon design.

His exposure to mechanical craftsmanship from an early age, his family's military tradition, formal education in engineering, and personal experiences as a soldier all contributed to his determination to improve upon existing firearm technology.

***

Johann Nicolaus von Dreyse, often referred to as the Father of Breech-loading Rifles, was an exceptional German gunsmith and inventor whose contributions revolutionized the world of firearms. Among his many accomplishments, Dreyse's pioneering rifling technology stands out as a crucial advancement in gun design. By introducing innovative rifling techniques and improving the accuracy, range, and reliability of firearms, Dreyse played a pivotal role in shaping modern weaponry.

Rifling refers to the spiral grooves carved inside the barrel of a firearm. These grooves cause the projectile to spin upon exiting the barrel, stabilizing its flight and enhancing accuracy. While rifling itself was not a new concept during Dreyse's time, he made significant improvements that greatly impacted gun design. One of Dreyse's notable contributions was his invention of progressive or gain twist rifling.

Traditional rifling had a constant twist rate along the entire length of the barrel. However, through meticulous experimentation and observation, Dreyse discovered that varying twist rates could optimize bullet stabilization at different stages of its journey down the barrel. By using progressive rifling with a slower initial twist near the breech and gradually increasing twists towards the muzzle end, Dreyse achieved superior bullet stabilization throughout its entire trajectory.

This innovation led to increased accuracy over longer distances and improved consistency in shot placement. Additionally, Dreyse implemented polygonal rifling in his designs – a departure from traditional rounded grooves – which further enhanced performance. Polygonal rifling featured multiple flat surfaces instead of rounded lands and grooves found in conventional barrels. The benefits were twofold: first, polygonal barrels were easier to clean due to reduced fouling; secondly, they provided increased muzzle velocity by reducing gas leakage around bullets during firing.

This resulted in greater energy transfer to the projectile, thus improving both range and stopping power. Furthermore, Dreyse's rifling advancements contributed to the development of his renowned needle gun. The needle gun, also known as the Dreyse rifle, was one of the first successful breech-loading rifles. It utilized a self-contained cartridge that integrated a pin-like firing mechanism, making it faster and more reliable than muzzle-loaded firearms.

The needle gun's success can be attributed in large part to Dreyse's rifling technology. The combination of progressive rifling and polygonal barrels significantly improved accuracy, making the needle gun highly effective even at extended ranges. Moreover, its breech-loading design allowed for faster reloading times compared to traditional muzzle-loaders, giving soldiers a significant advantage on the battlefield. Johann Nicolaus von

Dreyse's pioneering rifling technology forever changed the landscape of firearm design.

His innovative approach to barrel rifling led to increased accuracy, extended range capabilities, and improved reliability in firearms. By perfecting techniques such as progressive rifling and introducing polygonal barrels into his designs like the needle gun, Dreyse solidified his place as an influential figure in advancing gun technology.

***

The invention and subsequent development of firearms have played a crucial role in shaping the course of human history. Throughout centuries, various advancements were made to improve the effectiveness, reliability, and speed of firearms. One such revolutionary advancement was the introduction of breech-loading rifles by Johann Nicolaus von Dreyse, often referred to as the "Father of Breech-loading Rifles." Dreyse's inventions marked a significant turning point in firearms technology, bringing about a revolution that forever changed the battlefield.

Before delving into the impact of Dreyse's inventions, it is essential to understand the historical context surrounding flintlock rifles. Flintlocks were widely used during the 17th and 18th centuries and operated by igniting gunpowder through a spark created by striking a flint against steel. While effective for their time, flintlocks had several limitations that hindered their practicality on the battlefield. One major drawback was their relatively slow reloading process.

To reload a flintlock rifle, soldiers had to pour gunpowder down the muzzle, insert a bullet or ball, and then tamp it all down using a ramrod. This process required soldiers to stand up while reloading, leaving them vulnerable to enemy fire. Additionally, flintlocks were prone to misfires due to wet weather conditions or faulty sparks from worn-out flints. Dreyse's breakthrough came in 1836 when he patented his innovative breech-loading rifle design.

Unlike traditional muzzle-loaders like flintlocks, Dreyse's rifle allowed soldiers to load bullets directly into the breech or rear end of the barrel rather than through the muzzle. This new mechanism significantly reduced reloading time as soldiers no longer needed to stand up and go through an intricate loading process. Furthermore, Dreyse incorporated an integral needle firing system into his design. This system involved a needle-like firing pin that pierced the cartridge primer upon pulling the trigger, igniting the gunpowder and propelling the bullet forward.

This mechanism greatly improved reliability and eliminated misfires caused by wet weather conditions. The impact of Dreyse's inventions was nothing short of revolutionary. Breech-loading rifles provided soldiers with a distinct advantage on the battlefield, allowing them to fire multiple rounds in quick succession. The increased rate of fire drastically changed military tactics, as armies could now deliver concentrated volleys of fire, inflicting heavy casualties on their enemies.

Moreover, Dreyse's rifle design inspired further advancements in firearms technology. His introduction of self-contained cartridges laid the foundation for future developments in ammunition manufacturing and made reloading even faster and more efficient. The adoption of breech-loading rifles revolutionized warfare strategies worldwide. Armies quickly recognized their potential and began transitioning from muzzle-loaders to Dreyse-inspired designs. These advancements played a pivotal role in shaping conflicts such as the American Civil War, where breech-loading rifles proved decisive in battles like Gettysburg.

Johann Nicolaus von Dreyse's inventions had a profound impact on firearms technology and military tactics.

***

Johann Nicolaus von Dreyse, widely acknowledged as the Father of Breech-loading Rifles, faced numerous challenges throughout his journey of developing and patenting his revolutionary firearm innovations. From financial constraints to technical hurdles, Dreyse's determination and perseverance played a pivotal role in overcoming these obstacles.

One of the primary challenges that Dreyse encountered was the lack of financial resources. In the early stages of his career, he faced difficulties in securing funding for his projects. Undeterred by this setback, he utilized his meager savings to establish a small workshop where he could experiment with different designs and mechanisms. However, without substantial financial backing, progress was slow.

Another significant obstacle Dreyse faced was the technical complexity involved in developing an effective breech-loading mechanism. Unlike traditional muzzle-loading firearms prevalent at the time, breech-loading rifles required careful engineering to ensure reliability and safety. Dreyse spent countless hours experimenting with various designs before finally perfecting his needle-fire system—a mechanism that used a needle-shaped firing pin to ignite a percussion cap inside the cartridge.

However, even after successfully developing a functional rifle prototype, obtaining patents for his inventions proved challenging for Dreyse. The patent application process involved navigating complex legalities and bureaucratic red tape. Furthermore, there were instances where other inventors claimed prior art or attempted to replicate Dreyse's work without acknowledging him as the original innovator.

Despite these obstacles, Dreyse demonstrated remarkable resilience in protecting his intellectual property rights. He tirelessly fought legal battles against those who attempted to infringe upon his patents or discredit his contributions. His determination paid off when he eventually obtained several patents for his inventions across Europe—ensuring recognition for his groundbreaking work.

In addition to financial constraints and legal battles over patents, societal resistance also posed challenges for Dreyse. During the early 19th century, muzzle-loading rifles were deeply ingrained in military traditions, and the idea of adopting breech-loading technology was met with skepticism. Convincing military officials and conservative factions about the advantages of his invention required persistence and persuasive skills.

Dreyse recognized that proving the superiority of his breech-loading rifles was crucial to gaining acceptance. He conducted extensive testing, showcasing their accuracy, rapid reloading capabilities, and increased firepower compared to traditional muzzle-loaders. These demonstrations gradually won over skeptics and convinced military leaders to adopt his innovative firearms.

The struggles faced by Johann Nicolaus von Dreyse in developing and patenting his firearm innovations were formidable. Overcoming financial limitations, technical complexities, legal battles, and societal resistance demanded unwavering determination. However, Dreyse's perseverance eventually paid off as he secured patents for his inventions and revolutionized firearms technology forever. His contributions laid the foundation for modern breech-loading rifles that continue to shape warfare and hunting even today.

***

In the vast realm of firearms history, one name stands out as a true pioneer in the development of modern weaponry: Johann Nicolaus von Dreyse. Known as the "Father of Breech-loading Rifles," his most significant contribution to firearms technology was undoubtedly the invention of the needle gun. This groundbreaking creation revolutionized warfare and left an indelible mark on military tactics and strategy.

Before delving into the significance of Dreyse's needle gun, it is crucial to understand its basic mechanics. Unlike traditional muzzle-loading rifles, which required time-consuming reloading procedures, Dreyse's invention featured a breech-loading mechanism that allowed for faster and more efficient reloading. The key innovation lay in its use of a self-contained cartridge containing both propellant and bullet—a concept far ahead of its time.

One cannot overstate the impact that this breakthrough had on military tactics during the mid-19th century. The rapid reload time offered by Dreyse's needle gun enabled soldiers to fire multiple rounds within minutes, transforming battles into intense exchanges where firepower became paramount. This advantage was particularly evident during conflicts such as the Austro-Prussian War and later in Prussia's stunning victory over France in 1870–71.

The strategic implications were profound. Armies equipped with traditional muzzle-loaders were simply unable to match the sheer volume of fire produced by soldiers armed with Dreyse's needle guns. This led to a significant shift in battlefield dynamics, favoring those who embraced this new technology.

Moreover, Dreyse's invention marked a turning point in small arms design—ushering in an era where breech-loading mechanisms became standard features on rifles worldwide. The needle gun served as inspiration for subsequent developments such as bolt-action rifles, semi-automatic weapons, and eventually fully automatic firearms.

Beyond its immediate military applications, the needle gun also had far-reaching consequences for society as a whole. Its widespread adoption led to a fundamental reevaluation of warfare, emphasizing the importance of technological superiority and rapid firepower. This shift in mindset paved the way for further advancements in military technology throughout history.

Furthermore, Dreyse's needle gun played a crucial role in shaping the development of industrial manufacturing processes. The mass production required to meet the demands of armed forces worldwide pushed factories and assembly lines to new levels of efficiency. This transformative effect was not limited to firearms alone but extended to other industries seeking to replicate this newfound productivity.

Johann Nicolaus von Dreyse's invention of the needle gun marked a pivotal moment in firearms history and warfare itself. Its breech-loading mechanism revolutionized military tactics, shifting the balance of power on the battlefield. The widespread adoption of this technology transformed both small arms design and industrial manufacturing processes. Today, we recognize Dreyse as a visionary whose breakthrough creation forever changed the course of history.

***

Johann Nicolaus von Dreyse's revolutionary invention, the breech-loading rifle, had a profound impact on warfare during his time. The Prussian Army recognized the potential of this new firearm and swiftly adopted it, forever changing the way battles were fought. This subtopic will delve into the military adoption of Dreyse's rifles by the Prussian Army and explore their significant impact on warfare.

The Prussian Army was quick to recognize the advantages offered by Dreyse's breech-loading rifles. Prior to their introduction, muzzle-loaded muskets were used in battles, which required soldiers to expose themselves in order to reload after every shot. This cumbersome process made soldiers vulnerable to enemy fire and significantly slowed down the rate of fire. The breech-loading mechanism of Dreyse's rifles allowed soldiers to load ammunition from behind a protective cover without exposing themselves on the battlefield.

The adoption of Dreyse's rifles by the Prussian Army brought about a revolutionary change in tactics. With these new firearms, soldiers could now fire multiple rounds per minute compared to just a few shots with muzzle-loaders. This increased rate of fire gave them a significant advantage over their enemies during engagements, enabling them to suppress enemy forces more effectively.

The accuracy of Dreyse's rifles also played a crucial role in transforming warfare. These rifles were known for their superior precision compared to other contemporary firearms. The rifled barrels provided better stability and improved bullet trajectory, allowing soldiers to hit targets accurately at longer distances than ever before. This accuracy allowed for more effective long-range engagements, giving Prussian troops an upper hand against adversaries who still relied on outdated muzzle-loaders.

Furthermore, the adoption of breech-loading rifles led to changes in infantry formations and tactics employed by the Prussian Army. With increased firepower and accuracy at their disposal, the Prussian infantry was able to adopt more flexible and aggressive tactics on the battlefield. The traditional linear formations used in previous battles were no longer necessary, as soldiers armed with Dreyse's rifles could now effectively engage enemies from various positions and angles.

The impact of Dreyse's rifles was evident during the Prussian victories in key battles. Notably, during the Austro-Prussian War of 1866, the Prussian Army armed with breech-loading rifles decimated the Austrian forces that still relied on muzzle-loaders. The superior firepower and accuracy provided by Dreyse's invention played a pivotal role in securing these victories for the Prussians.

The adoption of Johann Nicolaus von Dreyse's breech-loading rifles by the Prussian Army revolutionized warfare during his time. These rifles provided increased rate of fire, improved accuracy, and changed infantry tactics significantly. Their impact was evident in key battles where Prussian forces armed with these innovative firearms outmatched their adversaries still reliant on muzzle-loaders. Johann Nicolaus von Dreyse truly earned his title as "the Father of Breech-loading Rifles" through his contributions to military technology and warfare.

***

Johann Nicolaus von Dreyse, renowned as the Father of Breech-loading Rifles, not only revolutionized firearms technology but also left a lasting manufacturing legacy through the establishment of his firearm factory. This factory would go on to play a pivotal role in shaping the future of German industry. Dreyse's journey began in 1814 when he started experimenting with firearm designs. His breakthrough came in 1824 with the creation of the needle gun, an innovative breech-loading rifle that significantly improved reloading speed and accuracy.

Recognizing its potential, Dreyse founded his own firearm factory in Sömmerda, Germany, in 1828. The establishment of Dreyse's firearm factory marked a turning point in German industrial history. It became one of the first large-scale manufacturing facilities to produce firearms using advanced techniques and machinery. Dreyse's meticulous approach to production included implementing standardized processes, quality control measures, and modern machinery—practices that were ahead of their time.

One key aspect of Dreyse's manufacturing legacy was his emphasis on interchangeable parts. By developing standardized components that could be easily replaced or repaired, he streamlined production and facilitated efficient assembly line operations. This innovation greatly increased productivity while reducing costs associated with custom-made parts. Dreyse also recognized the importance of skilled labor in achieving high-quality output. To ensure consistent workmanship, he introduced comprehensive training programs for his employees—a practice uncommon during that era.

By investing in their professional development, Dreyse fostered a workforce capable of delivering precision craftsmanship crucial for producing reliable firearms. The impact of Dreyse's manufacturing practices extended beyond his own factory walls. As other industries observed his successful methods, they began adopting similar techniques—resulting in an overall improvement in German industrial practices during the mid-19th century. Dreyse's factory became a center for technological advancement, attracting engineers, machinists, and inventors eager to contribute to the growing firearms industry.

This concentration of talent and expertise led to further innovations in manufacturing processes, including the development of more advanced machinery and tools. Furthermore, Dreyse's factory created employment opportunities for the local community, stimulating economic growth in Sömmerda and its surrounding regions. The success of his enterprise inspired other entrepreneurs to invest in manufacturing facilities across Germany, fostering industrialization and contributing to the nation's economic prosperity.

Dreyse's firearm factory remained at the forefront of German industry for several decades. Even after his death in

1867, his family continued running the business successfully. The legacy he left behind influenced not only firearm production but also various other sectors that adapted to his manufacturing principles. Johann Nicolaus von Dreyse's establishment of a firearm factory marked a significant milestone in German industrial history.

His emphasis on standardized production processes, interchangeable parts, skilled labor training programs, and technological advancements had a profound impact on German industry as a whole.

***

Johann Nicolaus von Dreyse, the renowned German gunsmith, revolutionized firearms with his invention of the breech-loading rifle in the mid-19th century. This groundbreaking innovation not only changed the way wars were fought but also had a profound impact on hunting and sport shooting. Dreyse's rifles quickly gained global recognition and influence, spreading to various countries and leaving an indelible mark on military strategies and technological advancements.

One of the key reasons behind the worldwide recognition of Dreyse's breech-loading rifles was their superior performance compared to muzzle-loading firearms. The ability to load from the breech meant that soldiers could reload faster, improving their rate of fire in battles. This advantage was particularly evident during conflicts such as the Second Schleswig War (1864) and the Austro-Prussian War (1866), where armies equipped with Dreyse's rifles achieved significant victories against opponents still reliant on muzzle-loaders.

The Prussian Army was among the first to adopt Dreyse's revolutionary weapon system, recognizing its potential for modern warfare. The success of Prussian forces armed with these rifles during decisive battles like Königgrätz solidified their reputation as a formidable military power. As neighboring countries witnessed Prussia's triumphs, they quickly sought to acquire Dreyse's technology or develop their own breech-loading systems based on his design principles.

Beyond Europe, Dreyse's influence extended far and wide. Countries such as Japan recognized the superiority of these rifles and initiated efforts to import or locally produce similar weapons. The adoption of breech-loaders by Japanese forces played a crucial role in modernizing their military capabilities during conflicts like the Boshin War (1868-1869) and Satsuma Rebellion (1877). Similarly, nations in South America, including Brazil and Argentina, turned to Dreyse's rifles as they sought to upgrade their arsenals.

Dreyse's breech-loading rifles also found their way into the hands of civilians and hunters worldwide. The improved accuracy and ease of reloading made them highly sought after for both sport shooting and hunting purposes. Sporting clubs in various countries, from the United States to Australia, began incorporating these rifles into their activities, contributing to their global popularity. The influence of Dreyse's invention extended beyond the battlefield and hunting grounds.

Other firearm manufacturers recognized the advantages of breech-loading technology, leading them to develop their own variations. However, many of these designs were inspired by Dreyse's innovations. This further solidified his status as the father of breech-loading rifles and ensured that his influence persisted even as other manufacturers entered the market. Johann Nicolaus von Dreyse's breech-loading rifle revolutionized firearms technology and left an indelible mark on military strategies around the world.

The superior performance of these weapons quickly gained recognition among armies seeking an edge in warfare. Additionally, hunters and sport shooters embraced these rifles for their enhanced accuracy and reloading efficiency.

***

Johann Nicolaus von Dreyse, a name that resonates with reverence and admiration in the world of firearms history. As the father of breech-loading rifles, his inventions revolutionized warfare and forever changed the course of firearm development. Today, we pay homage to this legendary figure, reflecting on his life and achievements that continue to shape the modern world.

Born on November 20, 1787, in Sömmerda, Germany, Johann Nicolaus von Dreyse possessed an innate curiosity and talent for engineering from an early age. His relentless pursuit of perfection led him to create numerous groundbreaking innovations throughout his career. However, it was his invention of the breech-loading rifle that would secure him a place among the greatest firearm designers in history.

The advent of breech-loading rifles brought about a paradigm shift in military tactics. Prior to Dreyse's invention, muzzle-loading firearms were prevalent but had limitations regarding reloading speed and safety. With his revolutionary design incorporating a self-contained cartridge loaded from the rear end of the barrel, soldiers could fire multiple rounds rapidly without needing to expose themselves by standing up or shifting positions.

Dreyse's most notable creation was the Needle Gun, introduced in 1841 as an official Prussian army firearm. This exceptional weapon allowed soldiers to load cartridges into the breech with ease using a paper cartridge containing powder and bullets wrapped together for efficient loading and firing. The Needle Gun became renowned for its accuracy and rapid rate of fire during conflicts such as the Austro-Prussian War (1866) and the Franco-Prussian War (1870-71).

Its influence extended beyond German borders as other nations recognized its superiority.

The enduring legacy left by Johann Nicolaus von Dreyse extends far beyond his remarkable inventions alone. His contributions transformed military strategies worldwide while also influencing the civilian firearms industry. The concept of breech-loading mechanisms paved the way for further advancements, eventually leading to modern semi-automatic and automatic firearms.

Moreover, Dreyse's innovations inspired subsequent generations of firearm designers to push boundaries and explore new possibilities. His engineering prowess set a precedent for excellence and innovation that continues to shape the field of firearm development today.

In honoring Johann Nicolaus von Dreyse's remarkable achievements, we recognize not only his technical genius but also his impact on society. His inventions revolutionized warfare, shaping the outcomes of battles fought during his time and beyond. They enabled soldiers to defend their nations more effectively while reducing casualties on both sides.

Today, as we look back at the life and contributions of Johann Nicolaus von Dreyse, we are reminded that legends are not merely remembered but honored for their enduring legacies. His name will forever be associated with innovation, excellence, and a revolutionary spirit that changed the course of firearms history. As we continue to enjoy the benefits of his groundbreaking inventions, let us pay tribute to this visionary figure whose brilliance continues to inspire generations in their pursuit of progress and excellence in firearm design.

Eliphalet Remington
1973-1861

Eliphalet Remington, born on October 28, 1793, in Suffield, Connecticut, was a remarkable individual whose passion for firearms shaped the history of American manufacturing. From humble beginnings on his family farm to founding one of the most iconic gun companies in the world, Remington's innovative spirit and dedication to quality revolutionized the firearms industry. As a young man, Eliphalet Remington became fascinated with rifles and their mechanics.

He observed that most existing guns were unreliable and lacked accuracy. Determined to create a superior firearm, he set up a small blacksmith shop on his family's farm in Ilion Gorge, New York. There he began experimenting with new designs and techniques to improve weapon performance. Remington's first breakthrough came in 1816 when he crafted his own flintlock rifle. Its exceptional accuracy quickly gained recognition among local hunters and marksmen.

Encouraged by this success, Remington established himself as a reputable gunsmith and continued refining his designs. In 1828, Eliphalet Remington introduced his first patented innovation – a rifle barrel with seven grooves twisted into it instead of the traditional smoothbore design. This spiral rifling greatly enhanced accuracy by imparting spin to the bullet as it exited the barrel. The success of this invention led to increased demand for Remington's rifles throughout America.

Driven by ambition and an unwavering commitment to quality craftsmanship, Remington expanded his operations beyond his small blacksmith shop. In 1845, he constructed a factory building that would become the foundation for what is now known as the iconic Remington Arms Company. Eliphalet Remington's entrepreneurial spirit didn't stop at manufacturing firearms; he also recognized the importance of marketing and distribution networks.

As railroads began connecting towns across America, Remington seized the opportunity to expand his customer base. He established partnerships with distributors and created a sales network that reached far beyond his immediate vicinity. This strategic move allowed Remington Arms to become a national brand, firmly establishing it as a leader in the firearms industry. Eliphalet Remington's impact extended beyond the success of his company.

He played a crucial role in shaping American history during times of conflict. During the Civil War, Remington Arms supplied rifles and ammunition to Union troops, contributing significantly to their military advantage. Today, more than two centuries after its founding, Remington Arms continues to produce firearms that uphold Eliphalet Remington's legacy of innovation and quality craftsmanship. The company remains one of the most recognized names in the industry and stands as a testament to Eliphalet Remington's pioneering spirit.

Eliphalet Remington's life was marked by an unwavering commitment to excellence in firearm manufacturing. From his humble beginnings on a family farm to founding one of America's most iconic gun companies, he revolutionized the industry through innovative designs and relentless pursuit of quality.

***

Eliphalet Remington, the visionary behind Remington Arms, started his career as a humble blacksmith in the early 19th century. Little did he know that his passion for crafting metal would lead him on a remarkable journey to become one of America's most influential gunsmiths. Born in 1793 in Suffield, Connecticut, Eliphalet grew up surrounded by the clinking sounds of hammers and anvils in his father's blacksmith shop.

Fascinated by the artistry of metalworking from an early age, he began honing his skills under his father's guidance. However, it was during a hunting trip with friends that Eliphalet realized there was room for improvement when it came to firearms. In those days, muskets were the primary firearms used for hunting and self-defense. Dissatisfied with their accuracy and reliability, Eliphalet saw an opportunity to combine his blacksmithing expertise with his passion for firearms.

Determined to create better weapons, he set out on a path that would forever change the history of American gun manufacturing. At just twenty-three years old, Eliphalet established a small workshop near Ilion Gorge in upstate New York. With limited resources but boundless determination, he began experimenting with different designs and techniques. Combining precision craftsmanship with innovative ideas, Eliphalet soon crafted his first rifle - a flintlock model known as the "Remington Model I."

Word of Eliphalet's exceptional craftsmanship spread like wildfire among hunters and frontiersmen who were eager to acquire more accurate and reliable rifles. As demand soared, he expanded his workshop and invested in modern machinery to streamline production while maintaining quality standards. Eliphalet's commitment to excellence paid off when the U.S government recognized Remington Arms as one of the most trusted suppliers during the War of 1812.

His rifles played a significant role in defending American soldiers and securing victory. This recognition further cemented his reputation as a skilled gunsmith. As technology advanced, Eliphalet embraced change and adapted his designs accordingly. He transitioned from flintlocks to percussion cap ignition systems, which greatly improved reliability and reduced misfires. He also introduced new rifling techniques that enhanced accuracy over longer distances.

Eliphalet's dedication to innovation and constant improvement laid the foundation for Remington Arms' success in the years to come. His rifles became synonymous with quality craftsmanship, precision, and durability, earning him a loyal customer base both in America and abroad. Today, Eliphalet Remington's legacy lives on through Remington Arms, which has evolved into a global leader in firearms manufacturing. The company continues to uphold his values of craftsmanship and innovation while adapting to modern demands.

The journey from blacksmith to gunsmith was not an easy one for Eliphalet Remington. It required unwavering dedication, relentless pursuit of perfection, and an unyielding belief in his own abilities.

***

Eliphalet Remington, a skilled blacksmith from upstate New York, revolutionized the firearm industry in the early 19th century through his relentless pursuit of innovation. His commitment to quality and craftsmanship led to the creation of firearms that were not only reliable but also highly accurate. Through his inventions and dedication, Remington set a new standard for gun manufacturing that continues to shape the industry today.

One of Eliphalet Remington's most significant contributions was the development of interchangeable parts for firearms. Prior to his innovations, guns were typically handmade and required extensive customization for each individual piece. However, Remington saw an opportunity to streamline production by creating standardized parts that could be easily assembled and replaced if necessary. This breakthrough allowed for faster manufacturing processes, reduced costs, and improved maintenance and repair procedures.

Remington's obsession with precision led him to refine his manufacturing techniques further. He introduced new machinery and tools into his factory, enabling him to produce parts with unparalleled accuracy and consistency. By utilizing advanced metalworking methods such as milling and grinding, he achieved tight tolerances previously unheard of in firearm production. This attention to detail resulted in firearms that were not only aesthetically pleasing but also performed flawlessly.

Another groundbreaking innovation credited to Eliphalet Remington is the creation of the first successful practical breech-loading rifle in America. The traditional muzzle-loading rifles required loading ammunition from the front end (muzzle) of the barrel which was a time-consuming process during combat or hunting situations. Recognizing this inefficiency, Remington developed a mechanism where ammunition could be loaded from the rear (breech) end instead.

This advancement significantly improved reloading speed, making his rifles highly sought after by both military and civilian users.

Furthermore, Remington's commitment to innovation extended to the ammunition itself. He worked tirelessly to develop more efficient and reliable cartridges, experimenting with different bullet designs, propellants, and casings. His efforts resulted in the creation of powerful cartridges that improved range, accuracy, and overall performance of firearms. These advancements not only enhanced the shooting experience but also had a significant impact on the development of modern ammunition used today.

Eliphalet Remington's revolutionary innovations transformed the firearm industry by introducing standardized manufacturing processes, improving precision in production, revolutionizing loading mechanisms, and advancing ammunition technology. His relentless pursuit of excellence set a new standard for quality that continues to be synonymous with Remington Arms even after two centuries. Today, Remington firearms are renowned worldwide for their reliability, accuracy, and durability—a testament to Eliphalet Remington's enduring legacy as a pioneering gunsmith and visionary entrepreneur.

***

Eliphalet Remington, a visionary gunsmith and entrepreneur, founded Remington Arms in 1816 with a steadfast commitment to producing quality firearms. His dedication to innovation and craftsmanship laid the foundation for what would become one of the most respected firearms manufacturers in history.

Eliphalet's journey began in Ilion, New York, where he honed his skills as a blacksmith. As a young man, he became fascinated with the art of gun making and recognized an opportunity to improve upon existing designs. Eliphalet believed that by combining precision engineering with superior materials, he could create firearms that were not only reliable but also highly accurate.

Driven by his passion for excellence, Eliphalet set out to build his first rifle. He meticulously crafted every component by hand, ensuring that each part met his exacting standards. The result was a firearm that surpassed all expectations - it was durable, precise, and incredibly reliable. This initial success fueled Eliphalet's determination to establish his own manufacturing company.

In 1816, at the age of 23, Eliphalet founded Remington Arms Company. He rented a small workshop and assembled a team of skilled craftsmen who shared his vision for producing top-quality firearms. Together, they began manufacturing rifles using state-of-the-art machinery and innovative techniques.

Eliphalet's commitment to excellence extended beyond the production process - he also recognized the importance of customer satisfaction. He established a reputation for outstanding customer service by providing prompt repairs and replacements for any firearm that failed to meet expectations. This dedication built trust among customers who knew they could rely on Remington Arms for superior products.

One key aspect of Eliphalet's vision was continuous improvement through innovation. He encouraged his team to experiment with new designs and technologies in order to push the boundaries of firearm performance. This commitment to innovation led to several groundbreaking advancements, including the development of the Remington Rolling Block action, which revolutionized single-shot rifles.

Eliphalet's passion for quality also extended beyond his own company. He actively supported and collaborated with other gunsmiths, sharing his knowledge and expertise to elevate the entire industry. His contributions to firearm design and manufacturing techniques were widely recognized, earning him a reputation as a pioneer in the field.

Today, Remington Arms continues to uphold Eliphalet's legacy by producing firearms of unparalleled quality. The company's dedication to precision engineering and innovation remains at the core of its operations. Each firearm is meticulously crafted using advanced technologies and materials while adhering to Eliphalet's original vision for excellence.

Eliphalet Remington's vision for quality firearms laid the foundation for Remington Arms' success story. His relentless pursuit of perfection established a standard that has endured for over two centuries. As one of America's oldest continuously operating manufacturers, Remington Arms continues to honor Eliphalet's legacy by providing shooters with reliable, accurate, and high-performance firearms that reflect his unwavering commitment to excellence.

***

Remington Arms, founded by Eliphalet Remington in 1816, has left an indelible mark on American firearms history. From its humble beginnings as a small blacksmith shop to becoming one of the most iconic and influential gun manufacturers in the United States, Remington Arms revolutionized the industry with its innovative designs, commitment to quality, and dedication to meeting the evolving needs of firearm enthusiasts.

One of the key contributions that Remington Arms made to American firearms history was its pioneering role in the development of mass production techniques. Eliphalet Remington's invention of a specialized lathe allowed for interchangeable parts to be produced with unprecedented precision. This breakthrough not only increased efficiency but also facilitated easier repairs and replacements for gun owners. The concept of interchangeable parts became a cornerstone in firearms manufacturing and influenced other industries as well.

Additionally, Remington Arms played a significant role in shaping American military history. During the Civil War era, it produced large quantities of rifles for both Union and Confederate troops. The Union army heavily relied on Remington's Model 1861 rifle musket, which offered superior accuracy and reliability compared to many contemporaries. The company's ability to meet such high demands during times of war solidified its reputation as a reliable provider of firearms.

The introduction of iconic firearm models further cemented Remington's influence on American firearms history. The creation of legendary rifles like the Model 700 bolt-action rifle and shotguns such as the Model 870 pump-action shotgun established Remington Arms as a leader in innovation and quality craftsmanship. These firearms not only gained popularity among sport shooters but also found favor among law enforcement agencies and military personnel worldwide.

Remington's impact extended beyond individual models; it actively contributed to advancements in ammunition technology. The introduction of smokeless powder cartridges revolutionized ballistics by offering higher velocities and reducing fouling issues compared to traditional black powder. Remington's commitment to research and development led to the creation of numerous cartridges, including the .22 Long Rifle, .30-06 Springfield, and .223 Remington. These cartridges became staples in the shooting community and continue to be widely used today.

Furthermore, Remington Arms played a vital role in shaping American culture through its influence on hunting and shooting sports. The company's firearms became synonymous with American heritage, capturing the imagination of generations of firearm enthusiasts. The popularity of Remington firearms helped fuel interest in hunting and shooting sports across the country, contributing to their growth as significant recreational activities.

Remington Arms' impact on American firearms history is undeniable. Through its innovative manufacturing techniques, contributions to military conflicts, introduction of iconic firearms models, advancements in ammunition technology, and influence on American culture, Remington Arms has left a lasting legacy. As one of America's oldest gun manufacturers still in operation today, it continues to shape the industry while honoring its rich heritage established by Eliphalet Remington over two centuries ago.

***

The legacy of Eliphalet Remington, the pioneering gunsmith and founder of Remington Arms, extends far beyond his innovative firearms designs. His company not only revolutionized the way guns were manufactured but also played a significant role in shaping the Industrial Revolution itself. Remington Arms set new standards in manufacturing through their advanced techniques, efficient production processes, and commitment to quality.

One of the key contributions of Remington Arms to the industrial landscape was their adoption of interchangeable parts. While this concept had been explored by others before, it was Eliphalet's son, Philo Remington, who perfected it. By manufacturing parts that were identical in dimension and shape, they could be easily replaced without requiring extensive hand-fitting or customization. This breakthrough significantly reduced production time and costs while increasing overall efficiency.

Remington Arms also embraced automation to streamline their manufacturing processes. In 1828, they became one of the first companies to introduce steam-powered machinery into their factory. This innovative approach allowed for faster and more accurate production compared to traditional hand methods. The introduction of these machines not only increased output but also set a new standard for industrial manufacturing across various sectors.

Additionally, Remington Arms prioritized quality control as a cornerstone of their operations. They implemented rigorous testing procedures to ensure that every firearm leaving their factory met strict standards for reliability and accuracy. This commitment to excellence earned them an impeccable reputation among shooters and military organizations alike.

The company's dedication to quality extended beyond just its finished products; they also invested heavily in employee training and development. Recognizing that skilled workers were essential for maintaining high production standards, Remington Arms established apprenticeship programs that provided aspiring gunsmiths with comprehensive training on advanced techniques and machinery operation.

Furthermore, Remington Arms' impact on American industry went beyond its immediate sphere—its success inspired other manufacturers to adopt similar practices. The company's innovative manufacturing methods, including interchangeable parts and automation, influenced industries far beyond firearms. Their techniques were eventually adopted by a wide range of sectors, from automotive to electronics, forever changing the face of manufacturing in the United States.

Remington Arms not only left an indelible mark on the firearms industry but also played a pivotal role in shaping the Industrial Revolution. Through their pioneering approaches to manufacturing, such as interchangeable parts and automation, they set new standards for efficiency and quality. Their commitment to excellence and dedication to training also paved the way for advancements across various industries.

Eliphalet Remington's legacy continues to inspire modern manufacturers striving for innovation and success in an ever-evolving industrial landscape.

***

Eliphalet Remington, a visionary gunsmith and founder of Remington Arms, left behind a legacy that continues to shape the firearms industry to this day. His pioneering designs revolutionized the way firearms were manufactured and played a significant role in shaping American history. From his humble beginnings in Ilion, New York, to his lasting impact on the world of firearms, Eliphalet Remington's legacy is one carved in steel.

One of Remington's most enduring contributions was his innovative approach to firearm manufacturing. In an era when guns were typically custom-made by individual craftsmen, he introduced the concept of standardized parts and interchangeable components. This breakthrough allowed for mass production, making firearms more affordable and accessible to a wider range of customers. The assembly line process he implemented became an industry standard and laid the foundation for modern manufacturing techniques.

Remington's designs also prioritized functionality and reliability without compromising on quality. He understood that a firearm needed to perform flawlessly under any circumstances, whether it was for hunting or self-defense. This commitment to excellence earned him a reputation for producing some of the most durable firearms available at the time.

One of Remington's most iconic creations was the Model 1858 revolver. This six-shot handgun featured a solid frame design that ensured durability even during intense use. Its reliability made it popular among soldiers during the American Civil War as well as civilians seeking personal protection on the frontier. The Model 1858 became synonymous with Remington Arms' commitment to quality craftsmanship.

Remington also left his mark on rifle design with innovations like his Rolling Block action system. This breakthrough mechanism allowed for faster reloading times while maintaining exceptional accuracy—a crucial advantage on battlefields or during hunting expeditions where split-second decisions mattered. The Rolling Block rifles became renowned for their precision and reliability, further establishing Remington Arms as a leader in the firearms industry.

Beyond his contributions to firearms technology, Remington's legacy extends to the impact he had on American history. His rifles played a significant role in the westward expansion, as settlers relied on them for protection and sustenance. The durability and reliability of Remington firearms became essential tools for survival on the frontier.

Today, Remington Arms continues to build upon Eliphalet Remington's legacy by incorporating modern advancements into their designs while staying true to his original vision. Their commitment to quality and innovation ensures that each firearm they produce upholds the standards set by their founder over two centuries ago.

Eliphalet Remington's pioneering designs and manufacturing techniques revolutionized the firearms industry. His commitment to quality, functionality, and reliability established a standard that still influences firearm manufacturing today. From his innovative approach to mass production to his iconic revolver and rifle designs, Eliphalet Remington's legacy is one that has stood the test of time—a testament to his enduring influence in shaping American history and the world of firearms.

***

Eliphalet Remington, the pioneering gunsmith and founder of Remington Arms, left behind a remarkable legacy that extended far beyond his lifetime. The company he established played a crucial role in shaping military history, especially during the tumultuous periods of the Civil War and the World Wars. With their innovative firearms and unwavering commitment to quality, Remington Arms became synonymous with reliability and precision on the battlefield.

During the Civil War (1861-1865), when America was torn apart by internal strife, Remington Arms emerged as a vital supplier of firearms to both Union and Confederate forces. Their rifles, particularly the legendary Model 1858 "New Model Army," were highly sought after for their accuracy and durability. These weapons revolutionized warfare by introducing breech-loading technology, allowing soldiers to reload quickly under intense fire.

The superior design of Remington rifles gave soldiers an edge on the battlefield, contributing significantly to various key battles such as Gettysburg and Antietam.

As conflicts escalated into World War I (1914-1918), Remington Arms once again stepped up its game. Recognizing the need for modernization in military weaponry, they developed groundbreaking innovations such as automatic rifles and machine guns. One notable example was the M1917 Enfield rifle

- an American version of a British design - which became a standard issue for American troops during this global conflict.

The M1917's reliability and firepower proved instrumental in turning the tide of battle against enemy forces.

The legacy continued into World War II (1939-1945) when Remington Arms made significant contributions to Allied victory. With increased production capacity due to wartime demands, they supplied millions of firearms that equipped soldiers across all fronts. The M1903 Springfield rifle served as a primary weapon for American troops during this time; its accuracy made it indispensable for snipers and marksmen. Additionally, Remington Arms played a crucial role in producing the M1 Garand, one of the most iconic rifles of the war.

The semi-automatic capabilities of this rifle gave American soldiers a distinct advantage over their adversaries.

Beyond rifles, Remington Arms also manufactured artillery shells and ammunition during both world wars. Their expertise in munitions production ensured a steady supply for the armed forces. This support was particularly critical during World War II when industrial output determined victory or defeat.

The dedication to quality that Eliphalet Remington instilled in his company resonated throughout history. The firearms produced by Remington Arms not only influenced military strategies but also shaped the lives of countless servicemen who relied on their weapons for survival and success on the battlefield.

From its humble beginnings as a small gunsmith shop, Remington Arms grew into an industry leader whose contributions to military history cannot be overstated. Through their innovative firearms and commitment to excellence, they played an integral role in shaping conflicts from the Civil War through both World Wars. Eliphalet Remington's pioneering spirit lives on through his company's enduring legacy - one that continues to impact military operations around the world.

**

Throughout its rich history, Remington Arms has consistently demonstrated its ability to adapt to evolving technologies and markets, ensuring its position as a leader in the firearms industry. Founded by Eliphalet Remington in 1816, the company has overcome numerous challenges and embraced innovative approaches to remain at the forefront of firearm manufacturing. This subtopic explores how Remington Arms successfully navigated changing landscapes and harnessed new technologies.

In its early years, Remington Arms established itself as a pioneer in gunsmithing by introducing handcrafted rifles renowned for their quality and accuracy. However, as industrialization gained momentum in the late 19th century, new production methods emerged that threatened traditional craftsmanship. Recognizing the need for change, Remington Arms transitioned from manual production to mechanized manufacturing processes. By embracing emerging technologies such as precision machinery and assembly lines, the company increased efficiency while maintaining high standards of craftsmanship.

The advent of smokeless powder and self-contained metallic cartridges during this era presented another significant technological shift for firearms manufacturers. Remington Arms was quick to embrace these innovations and adapted its production lines accordingly. The company's expertise allowed it to capitalize on these advancements by producing reliable firearms that were compatible with modern ammunition technology.

As markets evolved throughout the 20th century, so did consumer demands for different types of firearms. In response, Remington Arms diversified its product line to cater to a broader range of shooting enthusiasts. From hunting rifles to shotguns for sporting purposes and military-style rifles for law enforcement agencies, the company expanded its offerings while maintaining a commitment to quality craftsmanship.

The digital age brought yet another wave of change that required adaptation from firearm manufacturers like Remington Arms. With increasing connectivity and technological advancements in design software and manufacturing processes, there emerged opportunities for innovation in both product development and marketing strategies.

Remington Arms capitalized on these opportunities by investing in research and development to create cutting-edge firearms that met the demands of the modern shooter. The company utilized computer-aided design (CAD) and advanced manufacturing techniques to enhance precision, ergonomics, and overall performance. Moreover, Remington Arms engaged with its customers through digital platforms, leveraging social media and online communities to foster brand loyalty and gather valuable feedback.

Remington Arms' ability to adapt to evolving technologies and markets has been integral to its enduring success as a firearm manufacturer. From embracing mechanization during the Industrial Revolution to capitalizing on advancements in ammunition technology and diversifying its product line, the company has consistently demonstrated its commitment to innovation. By riding the waves of change rather than resisting them, Remington Arms remains a leader in the industry while upholding Eliphalet Remington's legacy of pioneering gunsmithing.

***

Eliphalet Remington was a visionary gunsmith who laid the foundation for one of the most iconic firearms manufacturers in American history. His legacy continues to shape the industry and inspire generations of gun enthusiasts, hunters, and collectors. Today, Remington Arms stands as a testament to his ingenuity and commitment to excellence.

One aspect of Eliphalet Remington's enduring legacy is his unwavering dedication to quality craftsmanship. From its humble beginnings in Ilion, New York, Remington Arms has maintained a reputation for producing firearms that are reliable, accurate, and durable. This commitment to quality has earned them the trust of military forces around the world and made their firearms highly sought after by civilians.

In addition to their dedication to craftsmanship, Remington Arms has also played a significant role in shaping firearm technology over the years. From introducing bolt-action rifles during World War I to developing innovative designs like the Model 700 bolt-action rifle – which remains one of their best-selling firearms today – they have consistently pushed boundaries and set new standards within the industry.

Furthermore, Eliphalet Remington's vision extended beyond manufacturing exceptional firearms; he also understood the importance of fostering strong relationships with customers. This customer-centric approach remains integral to Remington Arms' business philosophy today. They continue to engage with their loyal customer base through various initiatives such as educational programs, shooting events, and community outreach programs aimed at promoting responsible gun ownership.

The company's commitment to preserving its heritage is evident through its efforts in historical preservation. The iconic Ilion plant where it all began still stands today as a symbol of American manufacturing excellence. Visitors can tour this historic facility and witness firsthand how Eliphalet Remington's legacy lives on through every firearm that is produced there.

Moreover, Remington Arms actively supports museums dedicated to preserving firearm history. They contribute to exhibits and collections that showcase the evolution of firearms over the years, paying homage to Eliphalet Remington's pioneering contributions.

The modern-day legacy of Eliphalet Remington and his company also extends beyond the manufacturing plant and historical preservation efforts. Remington Arms continues to invest in research and development, striving to create cutting-edge firearms that meet the evolving needs of its customers. Their commitment to innovation ensures that they remain at the forefront of the industry, adapting to new technologies and market trends while staying true to their heritage.

Eliphalet Remington's legacy as a pioneering gunsmith lives on through Remington Arms. The company's unwavering commitment to quality craftsmanship, innovation, customer engagement, historical preservation, and research and development ensures that his vision continues to shape the industry for generations to come. As firearm enthusiasts explore the modern-day legacy of Eliphalet Remington, they are reminded of his enduring impact on American firearm history.

Oliver Winchester

1810-1880

In the annals of American history, there are few stories as captivating and inspiring as that of Oliver Winchester, a man who transformed himself from a humble shirtmaker into a prominent gun manufacturer. With determination, innovation, and an unwavering entrepreneurial spirit, Winchester's remarkable journey not only defined his own legacy but also left an indelible mark on the firearms industry.

Born on November 30, 1810, in Boston, Massachusetts, Oliver Fisher Winchester grew up in a world undergoing rapid industrialization. As a young man, he began his career as an apprentice in his father-in-law's shirt-manufacturing business. Although he showed promise in this trade and even became a partner in the company later, it was Winchester's relentless pursuit of new opportunities that would lead him down an entirely different path.

During the mid-19th century, America was experiencing unprecedented growth and expansion. This era marked the shift from muzzle-loading firearms to more technologically advanced repeating rifles. Recognizing this transformative trend within the industry, Winchester saw an opportunity for both personal success and contributing to American innovation.

Winchester's first foray into firearms came when he invested in the Volcanic Repeating Arms Company in 1855. The company held patents for lever-action repeating rifles that boasted superior firepower compared to their contemporaries. Sensing potential despite initial setbacks faced by Volcanic Arms Company due to financial difficulties and product design flaws, Winchester took over its assets after it went bankrupt in 1857.

Under his leadership and vision, Winchester reorganized the company as New Haven Arms Company. He worked tirelessly with talented engineers like Benjamin Tyler Henry to improve upon existing designs and develop innovative mechanisms that would revolutionize firearm manufacturing forever. It was during this period that Henry created the legendary Henry rifle – one of the first successful lever-action repeating rifles.

In 1866, Winchester renamed the company once again, this time as Winchester Repeating Arms Company. With an unrelenting commitment to quality and precision, he built a reputation for producing reliable firearms that became synonymous with the American West. The Winchester Model

1873, known as "The Gun That Won the West," became an iconic symbol of American frontier expansion.

While Oliver Winchester's success in the firearms industry was undeniable, his journey was not without its challenges. He faced fierce competition from other manufacturers and had to navigate through periods of economic downturns and societal upheavals like the Civil War. However, his unwavering dedication to excellence and continuous innovation allowed him to overcome these obstacles and solidify his position as one of America's most influential entrepreneurs.

As we delve deeper into Oliver Winchester's extraordinary story, we will witness how his relentless pursuit of success transformed a simple shirtmaker into a prominent figure in American industrial history. His remarkable journey serves as an inspiration for generations to come, reminding us that with passion, perseverance, and ingenuity, ordinary individuals can achieve extraordinary feats.

***

Oliver Winchester, a name synonymous with firearms and innovation, had humble beginnings that belied the remarkable journey he would embark upon. Born on November 30, 1810, in Boston, Massachusetts, Winchester initially found success in an unexpected industry before venturing into gun manufacturing. This subtopic explores the promising start of Oliver Winchester's early business ventures and sheds light on his entrepreneurial spirit.

In his youth, Oliver Winchester apprenticed as a shirtmaker under his father's tutelage. While the textile industry seemed to be his destined path, young Oliver possessed an innate curiosity and a desire to explore new opportunities. Sensing the changing economic landscape of mid-19th century America, he decided to take a leap of faith and invest in various business ventures.

One of Oliver's earliest endeavors was at a men's clothing store in Baltimore. He quickly recognized the importance of quality materials and craftsmanship when it came to clothing, principles that would later influence his approach to firearm manufacturing. Despite facing stiff competition from established tailors in the city, Winchester's attention to detail and commitment to customer satisfaction allowed him to carve out a niche for himself.

Winchester's keen eye for emerging markets soon led him into the realm of dry goods trade. Establishing successful partnerships with manufacturers across New England allowed him to build an extensive network of suppliers for various commodities such as fabrics, hardware, and tools. Through shrewd negotiation tactics and strong relationships with wholesalers, he achieved significant success in this field.

However, it was during one particular trip through Connecticut that fate intervened and forever altered the course of Oliver Winchester's life. En route to New York City for business purposes in 1855, he passed through New Haven - home to the renowned firearms manufacturer Volcanic Repeating Arms Company (VRAC). Intrigued by their innovative lever-action repeating rifles capable of firing multiple shots without reloading, Winchester seized the opportunity to purchase a significant portion of VRAC's shares.

This acquisition marked the turning point in Oliver Winchester's business career. Recognizing the immense potential of the firearm industry, he rebranded VRAC as the New Haven Arms Company and set out to revolutionize firearm manufacturing. Applying his meticulous attention to detail and commitment to excellence acquired during his earlier ventures, Winchester transformed New Haven Arms into a force to be reckoned with.

Oliver Winchester's early business ventures laid a solid foundation for his future success as a gun manufacturer. His experiences in tailoring and dry goods trade taught him invaluable lessons about quality control, supply chain management, and customer satisfaction. These principles would become the cornerstones of his approach to firearm production, propelling him toward becoming one of America's most influential entrepreneurs.

Oliver Winchester's journey from shirtmaker to gun manufacturer was not one of happenstance but rather an outcome of his entrepreneurial spirit and unwavering determination. The promising start he experienced through his early business ventures equipped him with vital skills that would ultimately shape the landscape of firearms manufacturing forever.

***

In the realm of American industry, few stories are as remarkable as that of Oliver Winchester. From humble beginnings as a shirtmaker, Winchester's journey led him to revolutionize the gun manufacturing industry. His relentless pursuit of innovation and commitment to quality not only transformed his own company but also left an indelible mark on American history.

Oliver Winchester's foray into the gun manufacturing industry began in 1857 when he acquired a controlling interest in the failing Volcanic Repeating Arms Company. Recognizing the potential for improvement, he set out to enhance their existing lever-action rifle design. This marked the beginning of his lifelong dedication to innovation.

Winchester's key breakthrough came in 1860 with the introduction of the Henry rifle, which featured a reliable and efficient lever-action mechanism. This firearm quickly gained popularity for its superior firepower and ease of use. However, it was only after the Civil War that Winchester truly seized upon his opportunity for success.

The years following the war were characterized by rapid industrialization and technological advancements in America. Oliver Winchester was at the forefront of this wave, embracing new production methods and machinery to increase efficiency and output without compromising quality. The introduction of interchangeable parts allowed for faster assembly and easier repairs, enabling mass production on an unprecedented scale.

Furthermore, Winchester recognized that diversification was crucial for long-term success. He expanded his product line beyond rifles to include shotguns, pistols, and even ammunition. This strategy not only broadened its customer base but also ensured that Winchester maintained a competitive edge over other manufacturers who focused solely on one type of firearm.

Another area where Oliver Winchester displayed remarkable innovation was marketing and salesmanship. He understood that appealing to consumer desires required more than just functional firearms; aesthetics played a significant role as well. To this end, he hired skilled engravers who transformed ordinary firearms into works of art with intricate designs and embellishments. This attention to detail not only enhanced the appeal of Winchester firearms but also solidified their reputation for quality.

Perhaps one of Winchester's most significant contributions to the gun manufacturing industry was his commitment to continuous improvement. He actively sought feedback from customers and used it to refine and enhance his designs. This iterative approach allowed Winchester firearms to consistently evolve, ensuring they remained at the forefront of technological advancements.

Oliver Winchester's remarkable journey from shirtmaker to gun manufacturer is a testament to the power of innovation. Through his relentless pursuit of improvement, he transformed an industry, leaving a lasting legacy that endures today. His commitment to quality, diversification, marketing prowess, and continuous improvement set a standard that many other manufacturers would strive to emulate in the years that followed. Oliver Winchester's name remains synonymous with excellence in gun manufacturing and serves as an inspiration for future innovators in any industry.

In the annals of American history, few firearms hold as much significance as the Winchester Rifle. Designed by Oliver Winchester, this extraordinary weapon revolutionized the world of firearms and cemented his name among the great inventors of his time. The creation of the Winchester Rifle marked a turning point in weaponry technology and forever changed the landscape of American firearms manufacturing.

Oliver Winchester's journey from a humble shirtmaker to a renowned gun manufacturer was an extraordinary one. Born in Boston in 1810, he initially ventured into business by establishing a successful men's clothing company. However, it was his acquisition of a controlling interest in the Volcanic Repeating Arms Company that opened up new horizons for him and set him on an entirely different path.

At that time, repeating rifles were still relatively new and faced significant challenges. They were notorious for their unreliability, complicated mechanisms, and limited ammunition capacity. Recognizing these flaws, Oliver set out to create a firearm that would address these issues and surpass all existing models.

Oliver's breakthrough came with the development of what would later be known as the Henry Rifle. Unlike its predecessors, this rifle featured a lever-action mechanism that allowed for rapid firing while maintaining accuracy and reliability. This innovation stemmed from Oliver's keen understanding of both engineering principles and market demands.

The Henry Rifle was chambered for .44 caliber rimfire cartridges, which significantly increased ammunition capacity compared to other rifles at that time. Furthermore, it had an internal magazine capable of holding up to fifteen rounds – an impressive feat considering its compact size. These advancements made it not only highly practical but also deadly efficient on the battlefield or during hunting expeditions.

The success of the Henry Rifle propelled Oliver Winchester into fame within firearm circles and laid the foundation for his future endeavors. Capitalizing on this achievement, he reorganized his company as The New Haven Arms Company in 1857. Under his leadership, the company flourished, emphasizing quality craftsmanship and continuous innovation.

However, it was in 1866 that Oliver Winchester's most iconic invention was born – the Winchester Model 1866 lever-action rifle. This rifle was built upon the successes of the Henry Rifle while incorporating additional improvements. The introduction of a loading gate allowed for easier reloading, and its stronger construction ensured better durability and longevity.

The Winchester Model 1866 became an instant sensation, earning a reputation as "the gun that won the West." Its reliability, ease of use, and remarkable accuracy made it a favorite among frontiersmen, law enforcement agencies, and even military units around the world.

Oliver Winchester's remarkable journey from shirtmaker to gun manufacturer culminated in his pioneering designs that forever changed firearm technology. His breakthrough invention, the Winchester Rifle series, represented a quantum leap in firearms manufacturing by embodying unparalleled reliability and superior performance. These rifles left an indelible mark on American history and solidified Oliver Winchester's legacy as one of the most influential figures in firearms development.

***

In the mid-19th century, firearms manufacturing underwent a transformative period with the introduction of lever-action rifles. At the forefront of this revolution stood Oliver Winchester, a businessman who transitioned from shirt-making to becoming one of the most influential figures in gun manufacturing history. Winchester's game-changing contributions to lever-action rifles not only revolutionized the industry but also left an indelible mark on American culture and warfare.

Oliver Winchester's journey into firearms began in 1855 when he became a major stockholder in the Volcanic Repeating Arms Company. Recognizing the immense potential of lever-action firearms, Winchester reorganized the struggling company and renamed it the New Haven Arms Company in 1857. Under his leadership, innovation flourished, leading to remarkable advancements that forever changed firearm design.

One of Winchester's most significant contributions was his role in perfecting Henry Deringer's original lever-action design. Lever-action rifles offered rapid-fire capabilities previously unseen in firearms, allowing users to fire multiple rounds without needing to manually reload after each shot. This innovation transformed weapons from single-shot muskets into efficient repeaters that could be employed for both hunting and warfare.

Winchester tirelessly worked towards refining and improving lever-action mechanisms, resulting in his company releasing iconic models such as the Henry Rifle and, later, its more famous successor, the Winchester Model 1866 "Yellow Boy." These rifles gained widespread popularity due to their reliability, durability, and ease of use. Lever actions became synonymous with American expansionism during this time period.

The impact of these rifles extended far beyond their functionality; they shaped American history itself. Lever actions played an instrumental role during America's westward expansion and were widely adopted by settlers, lawmen, outlaws, and Native American tribes alike. Their versatility allowed frontiersmen to hunt game for sustenance and defend themselves against various threats, while law enforcement agencies adopted them as standard-issue arms due to their firepower and ease of maintenance.

Furthermore, Winchester lever-action rifles left an indelible mark on warfare. During the American Civil War, both Union and Confederate soldiers recognized the value of these rapid-fire repeaters. The superior firepower provided by lever actions gave troops a significant advantage in battle, enabling them to deliver a higher volume of fire compared to muzzle-loading muskets. This advantage came at a time when traditional military tactics were being challenged by new technologies, solidifying Winchester's place in military history.

Oliver Winchester's contributions to lever-action rifles revolutionized the firearms industry and had far-reaching implications for American society. His relentless pursuit of innovation paved the way for future advancements in firearm design, ultimately shaping the way wars were fought and won. Today, lever-action rifles remain iconic symbols of American heritage and continue to be cherished by collectors and enthusiasts worldwide. The remarkable journey of Oliver Winchester from shirtmaker to gun manufacturer forever changed the trajectory of firearms history and left an enduring legacy that continues to resonate today.

***

Oliver Winchester's transformation from a humble shirtmaker to a dominant force in the firearms market is a testament to his entrepreneurial spirit, innovative mindset, and unwavering determination. Through strategic business decisions, relentless pursuit of quality, and a keen understanding of customer preferences, Winchester propelled his company to unparalleled heights. This subtopic explores the key milestones and strategies that led to Oliver Winchester's remarkable journey toward dominating the firearms market.

One of the pivotal moments in Oliver Winchester's career was when he acquired control of the Volcanic Repeating Arms Company in 1857. Recognizing the potential of their lever-action repeating rifle design, he invested heavily in refining its mechanism and enhancing its reliability. This marked Winchester's entry into firearm manufacturing and laid the foundation for his future success.

Winchester understood that mass production was crucial for commercial viability. He implemented cutting-edge manufacturing techniques, such as interchangeable parts, which allowed for faster assembly and repair processes. By streamlining production, he ensured consistency in quality while meeting increasing demand.

Additionally, Oliver Winchester recognized that innovation was necessary to stay ahead in an evolving market. His engineers developed new models like the Henry rifle, which featured a revolutionary magazine-fed system capable of holding 15 rounds—a significant improvement over competitors' offerings at the time. The Henry rifle quickly gained popularity among soldiers during the American Civil War due to its firepower advantage.

To sustain expansion and success, Winchester adopted an astute marketing strategy that focused on building trust with customers. He emphasized quality control by personally inspecting every firearm before it left his factory—an uncommon practice at that time. By associating his name with reliability and excellence, he won over consumers' confidence.

Furthermore, Oliver Winchester capitalized on opportunities presented by international markets. In 1866, he established sales branches in Europe to tap into growing demand abroad—an ambitious move given fierce competition from established European firearm manufacturers. Winchester's commitment to quality and relentless pursuit of excellence soon made his company a global brand, further solidifying his dominance in the firearms market.

Another key aspect of Winchester's success was his ability to adapt to changing customer preferences. He recognized the shift towards centerfire ammunition and leveraged this trend by introducing the Winchester Model 1873—a lever-action rifle that accepted centerfire cartridges. This model became known as "The Gun That Won the West" due to its popularity among settlers, lawmen, and cowboys alike.

Oliver Winchester's journey towards dominating the firearms market was not without challenges, but his unwavering commitment to innovation, quality control, and customer satisfaction propelled him to unparalleled success. By expanding his product line, embracing cutting-edge manufacturing techniques, building trust with customers through stringent quality control measures, and strategically targeting international markets, he established Winchester Repeating Arms Company as an iconic name in firearms history.

Oliver Winchester's legacy as a pioneer in the industry endures today through the continued success of his company and its enduring reputation for excellence.

***

The journey of Oliver Winchester from a humble shirtmaker to a renowned gun manufacturer is one filled with numerous challenges and remarkable triumphs. In an era marked by intense competition, Winchester faced numerous obstacles that tested his determination and innovation. Through perseverance, strategic decisions, and an unwavering commitment to quality, he not only survived but thrived in the fiercely competitive industry.

One of the primary challenges Winchester encountered was the dominance of established gun manufacturers during his early years in the business. Companies like Colt and Remington had already established themselves as leaders in the firearms industry, making it difficult for newcomers to gain a foothold. However, Winchester's unwavering belief in his own ideas and commitment to excellence pushed him forward.

To overcome this obstacle, Winchester focused on creating innovative designs that would set him apart from his competitors. He introduced several groundbreaking features in his rifles, such as the iconic lever-action mechanism that revolutionized firearm technology at the time. This invention not only differentiated Winchester's guns from those of other manufacturers but also appealed to customers seeking more efficient and reliable firearms.

Another significant challenge Winchester faced was financial instability. Starting a business requires substantial capital investment, which presents difficulties for someone transitioning from an entirely different industry. Despite facing financial setbacks early on, including bankruptcy after establishing his first company, C.S. Windust & Co., Winchester remained determined.

To secure financial stability for his ventures, he sought out investors who believed in his vision and were willing to provide the necessary funding. With their support, he founded the New Haven Arms Company, which later became known as Winchester Repeating Arms Company – an establishment that would become synonymous with quality firearms.

Winchester's triumph over financial hurdles can also be attributed to his astute decision-making skills. He recognized opportunities presented by government contracts during times of war when demand for firearms surged exponentially. By securing contracts with Union forces during the American Civil War, Winchester not only ensured financial stability for his company but also gained credibility and recognition as a reliable gun manufacturer.

Furthermore, Winchester's unwavering commitment to quality played a crucial role in overcoming challenges posed by competitors. While some manufacturers focused on mass production and cost-cutting measures, Winchester remained dedicated to producing firearms of superior quality. This commitment resonated with customers who valued reliability and durability in their firearms, establishing Winchester as a trusted brand.

Oliver Winchester's journey from shirtmaker to gun manufacturer was fraught with challenges that tested his resilience and determination. By overcoming obstacles such as fierce competition, financial instability, and the need to differentiate himself from established players in the industry, he ultimately achieved remarkable success. Through innovation, strategic decision-making, and an unwavering commitment to quality, Winchester carved out his place in history as one of the most influential figures in the firearms industry.

***

The legacy of quality and precision is deeply embedded in the history of Winchester firearms, making it a brand synonymous with excellence. From its humble beginnings as a shirt manufacturer, Oliver Winchester transformed his company into a leading gun manufacturer, leaving behind an enduring reputation for producing firearms of unparalleled craftsmanship.

Oliver Winchester's commitment to quality was evident from the early days of his foray into the firearms industry. Recognizing the need for precision and reliability in firearms, he invested heavily in advanced machinery and employed skilled craftsmen to ensure that every Winchester firearm met the highest standards. This dedication to superior craftsmanship quickly earned Winchester firearms a reputation for unrivaled quality.

One hallmark of Winchester's legacy is their continuous pursuit of innovation. Oliver Winchester understood the importance of staying ahead in an ever-evolving industry, constantly striving to improve upon existing designs and technologies. His commitment to innovation led to several groundbreaking advancements in firearm design, revolutionizing the industry.

One such innovation was the introduction of the lever-action repeating rifle, which became one of Winchester's most iconic products. This revolutionary design allowed shooters to fire multiple rounds rapidly without needing to reload manually after each shot. The lever-action mechanism also provided exceptional reliability and ease-of-use, further enhancing the overall shooting experience. The success and popularity of these lever-action rifles solidified Winchester's position as a leader in firearm manufacturing.

Winchester's commitment to precision manufacturing extended beyond their rifles. Their handgun offerings were equally renowned for their exceptional accuracy and reliability. From classic revolvers like the Model 1873 Single Action Army to modern semi-automatic pistols like the Model 1911, every Winchester handgun exemplified meticulous attention to detail and uncompromising performance.

Another aspect that contributes significantly to their enduring reputation is their unwavering dedication to customer satisfaction. Throughout its history, Winchester has maintained a strong connection with its customers by providing exceptional customer service and support. The company's commitment to standing behind their products and addressing any issues promptly has earned them unwavering loyalty from generations of firearm enthusiasts.

The legacy of quality and precision established by Oliver Winchester lives on today. Winchester firearms continue to be highly sought after by collectors, sports shooters, and hunters alike. Each firearm bearing the Winchester name is a testament to the brand's enduring reputation for excellence.

The legacy of quality and precision is a defining characteristic of Winchester firearms. Oliver Winchester's commitment to superior craftsmanship, innovation, and customer satisfaction laid the foundation for an enduring reputation that continues to shape the industry today. From their iconic lever-action rifles to their accurate handguns, Winchester firearms have become synonymous with excellence in the world of firearms manufacturing.

The remarkable journey from shirtmaker to gun manufacturer has left an indelible mark on history and solidified Winchester's place as a true legend in the industry.

Oliver Winchester, the renowned gun manufacturer, made significant contributions to American history not only through his firearms but also through his company's diversification and innovative business practices. While Winchester Repeating Arms Company is primarily known for its iconic rifles, it was their expansion into other industries that truly shaped their legacy.

One of the most notable ventures undertaken by the Winchester Repeating Arms Company was in the field of ammunition. Understanding the need for reliable and high-quality bullets to complement their rifles, Oliver Winchester established an ammunition manufacturing division within his company. This decision revolutionized the arms industry as it ensured customers had access to compatible ammunition specifically designed for Winchester rifles.

By controlling both ends of the supply chain, Winchester created a seamless experience for gun owners while solidifying his company's dominance in the market.

In addition to ammunition, Winchester diversified its product line into other areas such as bicycles. During the late 19th century, cycling experienced a boom in popularity, presenting an opportunity for entrepreneurial minds like Oliver Winchester. The company began producing bicycles under the name "Winchester Bicycles," leveraging their reputation for quality craftsmanship gained from firearm production. These bicycles quickly gained recognition and became popular among enthusiasts across America.

Furthermore, Winchester Repeating Arms Company made significant contributions to American architecture by introducing mass-produced standardized parts. Leveraging their expertise in precision engineering acquired from firearm manufacturing, they applied similar principles to architectural hardware such as door locks and hinges. This innovation allowed builders across America to construct homes more efficiently and economically while ensuring consistent quality throughout.

Another area where Oliver Winchester left an indelible mark was philanthropy. As his wealth grew alongside his company's success, he became increasingly involved in charitable endeavors aimed at improving society. One notable initiative he supported was establishing educational institutions such as Yale University's Sheffield Scientific School and Connecticut Agricultural College (now known as University of Connecticut). Winchester believed that education was the key to progress and prosperity, and his contributions played a crucial role in shaping the educational landscape of America.

Lastly, Winchester's company played a vital role in the development of American industry and manufacturing practices. By implementing innovative production techniques, such as interchangeable parts, assembly lines, and quality control standards, they set new benchmarks for efficiency and productivity. These practices not only influenced the arms industry but also had a ripple effect across various sectors of American manufacturing.

Oliver Winchester's company made significant contributions beyond guns that shaped American history. Through diversification into ammunition production, bicycles, architectural hardware, philanthropy, and advancements in manufacturing practices, Winchester Repeating Arms Company left an enduring legacy. Their innovative spirit and commitment to excellence continue to inspire generations even beyond their remarkable firearms.

***

Oliver Winchester, a man with humble beginnings as a shirtmaker, embarked on an extraordinary journey that would forever change the landscape of gun manufacturing. His relentless pursuit of innovation and unwavering commitment to quality propelled him to become one of the most influential figures in the industry. From his innovative lever-action repeating rifle to his visionary leadership, Winchester's legacy as a true pioneer in gun manufacturing will continue to inspire generations.

Throughout his career, Oliver Winchester demonstrated an unparalleled passion for excellence. He understood that success lay not only in creating exceptional firearms but also in providing customers with unparalleled reliability and accuracy. By introducing the lever-action repeating rifle, he revolutionized firearm technology and set new standards for efficiency and convenience. His commitment to quality was unwavering, as he constantly sought ways to improve his designs and ensure that every Winchester firearm lived up to its reputation.

Winchester's visionary leadership was another key factor in his remarkable journey. Under his guidance, the company expanded its product range and diversified into various markets. He recognized the importance of adapting to changing times and embraced technological advancements while still maintaining the core principles that made Winchester firearms exceptional.

By nurturing a culture of innovation within his company, he laid the foundation for continuous growth and success.

Beyond his business achievements, Oliver Winchester also played a crucial role in shaping American history. His rifles became synonymous with westward expansion during a time when settlers relied on them for protection and survival. The legendary Model 1873 became known as "The Gun That Won the West," solidifying Winchester's place as an iconic figure in American folklore.

Despite facing challenges along the way, such as fierce competition from other manufacturers and occasional setbacks, Oliver Winchester never wavered from his vision or compromised on quality. This steadfast determination allowed him not only to survive but thrive amidst adversity.

Today, over 150 years after the founding of the Winchester Repeating Arms Company, Oliver Winchester's legacy continues to thrive. His innovative designs and commitment to excellence have stood the test of time, making Winchester firearms highly sought after by collectors and enthusiasts around the world. The company's continued success is a testament to the enduring impact of his contributions to gun manufacturing.

In celebrating Oliver Winchester's remarkable journey, we must acknowledge that his achievements were not without controversy. The firearms industry has faced scrutiny over issues such as gun violence and responsible ownership. While it is important to recognize these concerns, it is equally crucial to separate them from Oliver Winchester's pioneering spirit and contributions to manufacturing technology.

Oliver Winchester's journey from shirtmaker to gun manufacturer exemplifies the power of vision, innovation, and unwavering dedication. His story serves as a reminder that with passion and perseverance, anyone can overcome adversity and achieve greatness. As we reflect on his remarkable accomplishments, let us celebrate Oliver Winchester as a true visionary whose legacy will continue to shape the world of gun manufacturing for generations to come.

Christian Sharps

1810-1875

In the annals of American history, certain figures loom large, their names etched in our collective consciousness. Yet, amidst these celebrated heroes, there are countless individuals who have made significant contributions but remain relatively unknown. Christian Sharps is one such figure—a remarkable inventor and innovator whose groundbreaking rifles revolutionized warfare and left an indelible mark on American history. Born on January 2, 1810, in Washington Township, New Jersey, Christian Sharps grew up in a modest household with a keen interest in mechanics and engineering.

His natural curiosity led him to experiment with various contraptions from an early age. However, it was his passion for firearms that would eventually set him on a path to reshape the very nature of warfare. Sharps' journey began when he apprenticed under renowned gunsmiths John Krider and Henry Deringer. This invaluable experience honed his skills and deepened his understanding of firearm mechanics.

Inspired by his mentors' expertise and driven by an insatiable desire for improvement, Sharps embarked on a tireless quest to develop a rifle that would surpass anything seen before. The turning point in Sharps' career came in 1848 when he patented his revolutionary design—the breech-loading rifle. Unlike traditional muzzle-loading rifles prevalent at the time, Sharps' creation featured a hinged breechblock that allowed the shooter to load cartridges from the rear of the barrel effortlessly.

This innovation drastically reduced reloading time—a crucial advantage on the battlefield—and marked a pivotal moment in military technology. Sharps' ingenuity did not stop there; he continued refining his designs with unwavering dedication. In 1851, he introduced an improved version—the Sharps Model 1851 Carbine—which gained popularity among soldiers and frontiersmen alike due to its accuracy and reliability. This compact rifle, renowned for its precision and durability, quickly became an iconic weapon of the American West.

Its reputation was further solidified during the American Civil War, where it found favor among Union troops who cherished its long-range capabilities. Beyond his contributions to firearms technology, Sharps' impact on American history extends to manufacturing practices. In 1851, he established the Sharps Rifle Manufacturing Company in Hartford, Connecticut—a visionary move that transformed firearms production from a craft industry into a large-scale manufacturing endeavor.

This pioneering approach allowed for greater efficiency and consistency in rifle production, enabling the widespread distribution of his inventions. Unfortunately, despite his groundbreaking achievements and lasting influence on warfare and industry, Christian Sharps remains largely unknown to the general public. His name is seldom mentioned alongside other prominent figures of his time. However, it is precisely this oversight that compels us to shed light on this unsung hero and ensure that his contributions are recognized and celebrated.

In this exploration of Christian Sharps' life and accomplishments, we will delve into the intricacies of his revolutionary rifles—examining their impact on military tactics—and explore how his manufacturing innovations shaped industrial practices in America.

***

In the annals of American history, there are few figures as instrumental in reshaping warfare as Christian Sharps. Born in 1810 in New Jersey, Sharps would go on to become a brilliant inventor and firearms designer, revolutionizing the world of rifles. His contributions forever altered the course of military tactics and strategy, making him an unsung hero whose impact cannot be overstated.

During the early 19th century, firearms were still relatively primitive compared to modern standards. Smoothbore muskets dominated the battlefield, with their limited range and accuracy. However, it was during this era that Sharps began experimenting with new designs that would lay the foundation for a new era of warfare.

Sharps' most significant contribution was his invention of the breech-loading rifle. Unlike traditional muzzle-loading firearms, which required reloading from the front end after each shot, Sharps' design allowed for loading from the rear or breech. This simple yet ingenious modification drastically increased both speed and efficiency on the battlefield.

With a breech-loading rifle like those designed by Sharps, soldiers could fire multiple shots within a minute instead of struggling to reload cumbersome muzzle-loaders. This gave them a decisive advantage over their adversaries who were still using older technology. The rapid rate of fire also meant that engagements could be won or lost within minutes rather than hours.

Furthermore, Sharps' rifles boasted remarkable accuracy over longer distances due to their rifled barrels. By introducing spiral grooves inside the barrel that imparted spin to bullets as they were fired, he greatly improved both range and precision compared to smoothbore muskets. This newfound accuracy enabled soldiers armed with Sharps rifles to engage targets at unprecedented distances.

The impact of these advancements became evident during various conflicts throughout American history, where soldiers armed with Sharps rifles proved victorious against their opponents using outdated weaponry. The Mexican-American War of 1846-1848 and the American Civil War from 1861-1865 were particularly transformative in showcasing the advantages of Sharps' rifles.

In addition to their military applications, Sharps rifles also found favor among hunters and frontiersmen. Their accuracy and reliability made them an indispensable tool for those traversing the untamed American wilderness. Sharps' designs were so highly regarded that they became synonymous with quality and precision, earning a reputation that endures to this day.

Despite his significant contributions, Christian Sharps remains relatively unknown in mainstream historical narratives. However, his impact on warfare cannot be overstated. By revolutionizing firearms technology through the invention of breech-loading rifles, he forever altered the course of military history. His weapons transformed battles into lightning-fast engagements, where accuracy and speed determined victory or defeat. Christian Sharps may be an unsung hero in American history, but his legacy as a visionary inventor lives on in the evolution of rifles and their profound impact on warfare.

***

In the annals of American history, there are countless unsung heroes whose contributions have shaped the nation in profound ways. One such figure is Christian Sharps, a remarkable inventor who revolutionized warfare with his groundbreaking rifles. Among his many inventions, the Sharps rifle stands out as an emblem of technological advancement and long-range accuracy. This article delves into the story of this unsung hero and explores how his innovative rifle forever reshaped warfare.

Born in 1810 in Washington, New Jersey, Christian Sharps possessed an innate curiosity and a passion for engineering from a young age. His inventive mind led him to work as an apprentice at a local gunsmith's shop, where he honed his skills and eventually became a master gunsmith himself. However, it was not until he established his own company that he truly unleashed his creative genius.

The year 1848 marked a turning point in American history with the discovery of gold in California. Thousands flocked to the West Coast in search of fortune, creating an urgent demand for reliable firearms capable of providing protection and sustenance during perilous journeys. Recognizing this need, Christian

Sharps designed a revolutionary rifle that would become synonymous with precision and reliability.

At the heart of Sharps' design was his patented breech-loading mechanism – an ingenious innovation that allowed soldiers to load their rifles from behind rather than through the muzzle. This breakthrough eliminated many time-consuming steps involved in traditional muzzle-loading firearms while significantly increasing reloading speed on the battlefield.

Furthermore, what truly set apart the Sharps rifle was its exceptional long-range accuracy. By incorporating a heavy octagonal barrel and implementing precise rifling techniques, Christian Sharps created a firearm capable of firing projectiles with remarkable precision over great distances – something previously unheard of on such a scale.

The practical implications were staggering that soldiers armed with Sharps rifles now had superior firepower against adversaries using conventional muzzle-loading weapons. In battles like the American Civil War, where long-range engagements were becoming increasingly common, the Sharps rifle provided a distinct advantage for those wielding it.

The Sharps rifle's long-range accuracy also found applications beyond warfare. Hunters and frontiersmen quickly recognized its potential, as it allowed them to efficiently engage targets from distances that were previously inaccessible. The rifle became synonymous with the American West, embodying the spirit of adventure and self-reliance that characterized those who explored and settled the vast frontier.

Christian Sharps' invention not only reshaped warfare but also left a lasting impact on American culture. The revolutionary breech-loading mechanism and exceptional long-range accuracy of his rifles set a new standard for firearms technology. Moreover, his innovative designs paved the way for future advancements in firearm development.

Despite his profound influence on American history, Christian Sharps remains relatively unknown to many today. However, by recognizing his contributions and understanding the significance of his inventions – particularly the Sharps rifle – we can truly appreciate how this unsung hero forever changed warfare and shaped an entire era in American history.

***

In the annals of American history, there are unsung heroes whose contributions have reshaped warfare and left an indelible mark on society. One such hero is Christian Sharps, a brilliant inventor who revolutionized marksmanship with his innovative rifles. Through his relentless pursuit of perfection and unwavering commitment to precision shooting, Sharps redefined the art of marksmanship and forever changed the face of warfare.

Christian Sharps was born in Washington Township, New Jersey in 1810. From an early age, he displayed a keen interest in firearms and their mechanisms. His natural aptitude for engineering led him to apprentice with renowned gunsmiths, where he honed his skills and developed a deep understanding of the intricacies involved in building accurate firearms.

It was during his time as a master gunsmith that Sharps conceived the idea for what would become his most influential invention – the Sharps rifle. Unlike its predecessors, this firearm featured a unique breech-loading mechanism that allowed for faster reloading times and increased accuracy. The innovative design incorporated a sliding block that sealed the chamber tightly when fired, preventing gas leakage and ensuring consistent performance.

The impact of Sharps' invention on marksmanship cannot be overstated. Prior to his rifles, soldiers relied on muskets that were slow to load and had limited range and accuracy. With the introduction of breech-loading rifles like those designed by Sharps, soldiers gained a significant advantage on the battlefield. The ability to reload quickly meant more shots could be fired per minute while maintaining accuracy over longer distances.

One notable example showcasing Christian Sharps' impact occurred during the American Civil War. Union soldiers armed with his rifles proved formidable adversaries against Confederate forces armed with traditional muskets. The superior range and accuracy offered by these weapons allowed Union troops to engage enemy forces from afar while minimizing their own exposure to danger. This advantage played a crucial role in several key battles, ultimately contributing to the Union's victory.

Beyond the battlefield, Sharps' rifles found widespread use among hunters and frontiersmen during the westward expansion of the United States. The rifles' precision shooting capabilities made them ideal for hunting game from a distance, allowing hunters to provide sustenance for their families more efficiently and safely.

Christian Sharps' impact on marksmanship cannot be overstated. His innovative rifle designs revolutionized warfare, providing soldiers with a distinct advantage on the battlefield. The precision shooting capabilities of his firearms reshaped military tactics and strategies, leading to significant victories in key conflicts throughout American history. Furthermore, his rifles found practical applications outside of warfare, becoming invaluable tools for hunters and frontiersmen alike.

Christian Sharps was truly an unsung hero whose contributions forever changed the face of marksmanship and left an enduring legacy in American history.

***

In the annals of American history, few names shine brighter than Christian Sharps, an unsung hero whose revolutionary rifle played a pivotal role in reshaping warfare during the 19th century. The birth of the Sharps rifle marked a turning point in military tactics and forever altered the course of American combat. Christian Sharps, a skilled gunsmith and inventor, dedicated his life to perfecting firearms that would provide soldiers with a distinct advantage on the battlefield.

His determination paid off when he introduced his masterpiece – the Sharps rifle – in 1848. At its core, what set the Sharps rifle apart from its contemporaries was its innovative breech-loading mechanism. Unlike traditional muskets that required time-consuming muzzle loading, this new design allowed soldiers to load their rifles from the rear by sliding a cartridge into the breech. This breakthrough eliminated much of the downtime previously associated with reloading firearms, giving troops an undeniable advantage over their adversaries.

The introduction of breech-loading rifles revolutionized warfare by greatly increasing both accuracy and rate of fire. With traditional muskets, soldiers were lucky to fire two rounds per minute accurately. In contrast, an experienced soldier armed with a Sharps rifle could fire up to ten rounds per minute while maintaining exceptional accuracy. This unprecedented firepower placed unprecedented pressure on enemy forces and fundamentally changed battle strategies.

Moreover, the design of this firearm incorporated several other ingenious features that further enhanced its effectiveness on the battlefield. The falling block action not only provided excellent stability but also ensured reliable ignition regardless of weather conditions or fouling within the barrel. Additionally, due to its unique percussion system and self-contained cartridges, misfires were significantly reduced compared to other contemporary rifles.

The versatility offered by these advancements made it possible for troops armed with Sharps rifles to excel in various combat situations. From long-range engagements to close-quarters skirmishes, the Sharps rifle proved to be an adaptable and lethal weapon. Its outstanding accuracy, combined with its extended range, allowed soldiers to engage targets beyond the reach of traditional muskets, effectively outranging enemy forces.

The impact of the Sharps rifle was perhaps most evident during the American Civil War. Both Union and Confederate soldiers recognized its superiority and eagerly sought to acquire these game-changing firearms. The rifle's effectiveness in long-range engagements gave sharpshooters equipped with Sharps rifles a significant advantage on the battlefield. Notably, during the Battle of Gettysburg in 1863, Union sharpshooters armed with these rifles played a crucial role in turning the tide of battle.

The birth of Christian Sharps' revolutionary rifle marked a turning point in American military history. By introducing his innovative breech-loading mechanism and incorporating several other ingenious features into his design, Sharps single-handedly changed the face of warfare. The unmatched accuracy, increased rate of fire, and extended range provided by this firearm undoubtedly reshaped battle strategies and left an indelible mark on American history.

***

Christian Sharps, a name often overlooked in the annals of American history, played a pivotal role in reshaping warfare with his revolutionary rifles. Born in Washington, New Jersey in 1810, Sharps displayed an innate curiosity and aptitude for mechanics from an early age. His journey from invention to innovation would forever change the landscape of firearms and their impact on warfare.

Sharps' journey began with his invention of the "Sharps Rifle" in the mid-1840s. This breakthrough design featured a breech-loading mechanism that allowed for faster reloading and increased accuracy compared to muzzle-loading firearms of the time. The rifle's innovative design was simple yet effective, enabling soldiers to load cartridges directly into the chamber instead of having to pour powder down the barrel.

This advancement significantly reduced reloading time and greatly improved efficiency on the battlefield.

The success of Sharps' initial invention led him to establish his own manufacturing company, which he aptly named "C. Sharps & Company." With his newfound resources, he continued refining his designs and introducing further innovations that would forever transform rifles. One such innovation was the incorporation of a metallic cartridge case instead of paper or linen ones used by other contemporary firearms manufacturers.

Sharps recognized that metallic cartridge cases provided superior protection for propellant charges while also simplifying loading procedures for soldiers. Additionally, these cartridges were more durable and less prone to moisture damage than their paper counterparts. This breakthrough had far-reaching implications not only for military use but also for civilian applications such as hunting and sport shooting.

Another significant contribution by Christian Sharps was his development of telescopic sights or scopes for rifles. While not entirely new at the time, Sharps refined existing designs and made them more accessible by incorporating them into his rifle models. The addition of telescopic sights vastly improved accuracy over long distances, giving shooters a clear advantage in both military and civilian contexts.

Sharps' continuous dedication to innovation and improvement extended beyond his initial inventions. He sought feedback from soldiers, hunters, and marksmen, incorporating their suggestions into his designs. This collaborative approach ensured that his rifles were not only reliable but also user-friendly, catering to the needs of the individuals who relied on them in various scenarios.

The impact of Christian Sharps' inventions and innovations on warfare cannot be overstated. His breech-loading mechanism, metallic cartridges, and telescopic sights revolutionized the capabilities of firearms during a critical period in American history. These advancements allowed soldiers to reload faster, shoot with greater accuracy, and engage enemies from longer distances than ever before.

Christian Sharps may be an unsung hero in American history, but his contributions to firearms technology have left an indelible mark on warfare and hunting for generations to come. His relentless pursuit of improvement transformed rifles forever, shaping the way battles were fought and ultimately influencing the course of history itself.

***

The advent of the Sharps rifle in the mid-19th century revolutionized warfare on American battlefields. Designed by Christian Sharps, this innovative firearm played a crucial role in reshaping military strategies and tactics, ultimately leading to significant changes in the way wars were fought. From its exceptional accuracy and range to its rapid reloading capabilities, the Sharps rifle proved to be a game-changer that left an indelible mark on American history.

One of the most notable features of the Sharps rifle was its remarkable accuracy. Unlike earlier muskets or smoothbore firearms that were notoriously inaccurate beyond short distances, the rifling technology employed in Sharps rifles provided unrivaled precision. This allowed soldiers to engage targets at much longer ranges with confidence, giving them a significant advantage over their adversaries. The increased accuracy not only improved individual marksmanship but also enabled more efficient targeting of enemy officers and leaders during engagements.

Another key aspect that set the Sharps rifle apart was its unique breech-loading mechanism. Prior to its introduction, muzzle-loading rifles were prevalent, requiring soldiers to load ammunition from the muzzle end of their guns. However, Christian Sharps invented a breech-loading system that allowed cartridges to be loaded directly into the chamber at the rear of the barrel. This breakthrough innovation drastically reduced reloading time compared to muzzle-loaders and enabled soldiers to fire multiple shots rapidly without compromising accuracy or safety.

The rapid reloading capabilities of the Sharps rifle played a crucial role in changing military strategies during battles. Previously, armies relied heavily on massed volleys and slow reloading rates for effective firepower on the battlefield. With its breech-loading mechanism and self-contained cartridges, soldiers armed with Sharps rifles could reload swiftly while remaining behind cover or prone positions—a tactical advantage that disrupted traditional linear formations and forced opponents to adapt their strategies accordingly.

Furthermore, due to its range and accuracy, the Sharps rifle allowed for new tactics to emerge on the battlefield. Sharpshooters armed with these rifles could engage targets from concealed positions, rendering traditional fortifications less effective. They could pick off key enemy personnel, disrupt supply lines, and create chaos among enemy ranks without exposing themselves to undue risks. This introduced a new dimension of asymmetrical warfare into conflicts, where small units armed with Sharps rifles could have an outsized impact on the outcome of battles.

The influence of the Sharps rifle extended beyond its immediate military applications. Its success in combat led to widespread adoption in civilian life as well, particularly in the Western frontier. Frontiersmen, hunters, and explorers embraced this reliable and accurate firearm for self-defense and hunting purposes, contributing to the popularization of breech-loading technology throughout American society. Christian Sharps' innovative rifle transformed American warfare by revolutionizing accuracy and reloading capabilities on the battlefield.

The exceptional precision offered by the Sharps rifle enabled soldiers to engage targets at longer ranges with deadly effectiveness. Moreover, its breech-loading mechanism allowed for rapid reloading rates that disrupted traditional tactics and formations.

Christian Sharps is a name that has faded into the annals of history, yet his impact on American warfare and technological advancements cannot be overstated. Despite being an unsung hero, Sharps reshaped the course of warfare with his revolutionary rifles, leaving behind a forgotten legacy that deserves recognition. Born in Washington, New Jersey in 1810, Christian Sharps displayed an innate talent for engineering from an early age.

His passion for firearms led him to become an apprentice at the Harpers Ferry Armory in Virginia. During his time there, he honed his skills and developed a keen understanding of weapon mechanics. Sharps' breakthrough came in 1848 when he patented his first rifle design – the Sharps Rifle. This innovative firearm featured a sliding breechblock mechanism that improved loading speed and accuracy compared to traditional muzzle-loading rifles.

The design also allowed for easier cleaning and maintenance, making it highly practical for soldiers on the battlefield. The significance of Sharps' contribution lay not only in his rifle's technical advancements but also in its impact on military tactics. Prior to the introduction of the Sharps Rifle, infantrymen relied heavily on cumbersome muskets with limited range and accuracy. With its increased firepower and long-range capabilities, the new rifle revolutionized combat strategies by enabling soldiers to engage enemies from greater distances effectively.

During the American Civil War (1861-1865), Sharps Rifles became widely adopted by both Union and Confederate forces. Union troops armed with these weapons gained a significant advantage over their opponents due to their superior range and accuracy. This led to decisive victories in battles such as Gettysburg and Antietam. The Confederate army recognized this threat promptly and began procuring captured or privately purchased Sharps Rifles themselves.

Beyond its influence on warfare during the Civil War era, Christian Sharps' invention continued to shape American history. The rifles played a pivotal role in the westward expansion, as settlers and frontiersmen relied on them for protection and survival. From the famous "Buffalo Soldiers" to legendary figures like Buffalo Bill Cody, Sharps Rifles became synonymous with the American frontier. Despite its undeniable impact, Sharps' legacy gradually faded from public memory as new inventions and technological advancements overshadowed his achievements.

Over time, other firearms, such as repeating rifles, gained prominence, relegating the Sharps Rifle to a historical relic. However, it is crucial to recognize Christian Sharps' contribution not only as an inventor but also as a pioneer who revolutionized warfare. His innovative design forever changed military tactics and set the stage for future advancements in firearms technology. By appreciating his forgotten legacy, we gain a deeper understanding of how individuals like Sharps shaped American history through their ingenuity and perseverance.

Christian Sharps' rifles transformed warfare during the Civil War and left an indelible mark on American history. While his name may have been forgotten by many, his contributions should not be overlooked or diminished.

Christian Sharps, an unsung hero in American history, reshaped warfare with his revolutionary rifles. However, the impact of his inventions extended far beyond the battlefield. Sharps' rifles not only transformed military tactics but also played a crucial role in shaping various aspects of American society. From westward expansion to technological advancements and even cultural changes, the lasting impact of Christian Sharps' rifles is undeniable.

One significant consequence of Sharps' rifles was their influence on westward expansion in the United States. As pioneers pushed further into uncharted territories, they relied heavily on firearms for survival and protection. The accuracy and range offered by Sharps' rifles were unparalleled at the time, providing settlers with a distinct advantage over indigenous populations and wildlife alike. This advantage allowed pioneers to secure their claims more effectively and establish settlements that would eventually grow into thriving towns and cities.

Moreover, Christian Sharps' rifles played a vital role in technological advancements during this era. The precision engineering required to create these firearms was groundbreaking for its time. As demand for these rifles increased, factories developed new manufacturing processes to meet it efficiently. This led to advancements not only in firearms manufacturing but also in other industries that adopted similar techniques.

In addition to facilitating westward expansion and fueling technological progress, Sharps' rifles also had a profound impact on American culture. Firearms became an integral part of the American identity during this period, symbolizing independence, self-reliance, and rugged individualism. The popularity of sharps-style breech-loading rifles helped solidify this connection between guns and American ideals.

Furthermore, Christian Sharps' inventive spirit set a precedent for future inventors and innovators across various fields. His ability to identify shortcomings in existing technology and develop innovative solutions inspired others to think outside the box. This mindset would go on to shape countless industries beyond firearms manufacturing, fostering a culture of innovation that remains a defining characteristic of American society today.

The legacy of Christian Sharps' rifles can also be seen in the continued advancements in firearms technology. His designs laid the foundation for future developments, such as cartridge-based systems and repeating rifles. These innovations not only revolutionized warfare but also influenced hunting practices, law enforcement tactics, and self-defense strategies.

Christian Sharps' rifles had a profound and lasting impact on American society beyond the battlefield. Their influence is evident in westward expansion, technological advancements, cultural changes, and the fostering of innovation. The legacy of Sharps' inventive spirit can still be felt today in various industries and continues to shape American society's values and ideals. As an unsung hero in American history, Christian Sharps deserves recognition for reshaping warfare and leaving an indelible mark on our nation's development.

***

Christian Sharps, an unsung hero in American history, holds a remarkable legacy for his contributions to reshaping warfare with his rifles. From his humble beginnings as an apprentice gunsmith to establishing the renowned Sharps Rifle Manufacturing Company, Sharps played a pivotal role in revolutionizing firearms technology and leaving an indelible mark on American history. As we conclude our exploration of his life and achievements, it is crucial to honor his exceptional legacy and recognize the profound impact he had on shaping America.

First and foremost, Christian Sharps' ingenuity and innovation transformed the landscape of warfare during the mid-19th century. His revolutionary designs, such as the breech-loading mechanism incorporated into his rifles, introduced rapid-fire capabilities that rendered traditional muzzle-loading firearms obsolete. By enabling soldiers to reload their weapons from behind cover quickly, he significantly enhanced their efficiency on the battlefield. The widespread adoption of Sharps rifles during the Civil War exemplifies how this technological advancement revolutionized military tactics, ultimately influencing the outcome of many battles.

Furthermore, Christian Sharps' contributions extended beyond military applications. His inventions also played a vital role in shaping America's westward expansion during the mid-1800s. The ruggedness and reliability of his rifles made them well-suited for frontiersmen and pioneers traversing treacherous terrains filled with hostile Native American tribes. With their superior accuracy and range compared to traditional muskets or pistols, these weapons provided a significant advantage for those seeking safety and success in uncharted territories.

Moreover, Christian Sharps' entrepreneurial spirit deserves recognition as he successfully established one of America's most prominent firearm manufacturing companies. Through dedication and unwavering commitment to quality craftsmanship, he built a brand that became synonymous with innovation and excellence. The impact of his company extended far beyond its initial success; it laid the foundation for future firearm manufacturers and inspired countless inventors to push the boundaries of what was possible in weaponry design.

Despite his immense contributions, Christian Sharps remains relatively unknown to many Americans. It is, therefore, crucial that we honor his remarkable legacy and ensure his story is told and remembered for generations to come. By recognizing Sharps as a true pioneer in firearms technology, we not only pay homage to an unsung hero but also gain a deeper understanding of the profound impact individuals can have on shaping history.

Christian Sharps' role in reshaping warfare with his rifles cannot be overstated. His innovative designs revolutionized military tactics during the Civil War while also enabling America's westward expansion. The establishment of the Sharps Rifle Manufacturing Company further solidified his lasting impact on American history. By honoring his remarkable legacy, we ensure that future generations recognize the invaluable contributions of this unsung hero who changed the course of American warfare forever.

Samuel Colt
1814-1862

In the annals of American history, few names stand as tall as that of Samuel Colt. Born on July 19, 1814, in Hartford, Connecticut, Colt would go on to become one of the most influential figures in firearms manufacturing and a pioneer in industrial innovation. His revolutionary invention, the Colt revolver, forever changed the landscape of warfare and transformed the course of American history.

This subtopic delves into the remarkable life and enduring legacy of Samuel Colt.

From an early age, it was evident that Samuel Colt possessed an innate curiosity and a relentless drive for innovation. As a young boy growing up in Hartford, he displayed an extraordinary knack for mechanics and engineering. Fascinated by machinery since childhood, he would often dismantle household objects just to understand their inner workings. This precocious curiosity laid the foundation for his future accomplishments.

Colt's ingenuity first took shape during his voyage to India as a young sailor at the age of sixteen. Inspired by his experiences aboard ship, where he observed how multiple barrels could be rotated effortlessly with just one trigger mechanism on deck cannons, it sparked an idea that would change firearms forever. Upon returning to America in 1832 after several years at sea, he set out to bring this concept to life.

After much experimentation and refinement over several years, Samuel Colt patented his first revolver design in 1836 – a milestone moment that marked the birth of his illustrious career as a gun inventor. The unique feature of this firearm was its revolving cylinder capable of holding multiple bullets – a groundbreaking innovation that allowed rapid-fire capability without needing to reload after each shot.

Colt's invention quickly gained recognition not only for its functionality but also for its reliability and simplicity compared to other existing firearms at that time. Recognizing its potential military applications, he tirelessly pursued contracts with various governments across continents while also catering to civilian demand. His efforts paid off handsomely when the Texas Rangers adopted Colt's revolvers, paving the way for widespread acceptance and commercial success.

The impact of Samuel Colt's invention extended far beyond just firearms manufacturing. His technological advancements in mass production techniques played a pivotal role in shaping

American industrialization during the mid-19th century. With his innovative assembly line and interchangeable parts, Colt revolutionized manufacturing processes, becoming a trailblazer for future industrial giants like Henry Ford.

Samuel Colt's relentless pursuit of excellence led to numerous patents and improvements over the years. He not only transformed the world of firearms but also left an indelible mark on American culture and history. Today, his legacy lives on as his name remains synonymous with ingenuity, innovation, and perseverance.

In this subtopic about Samuel Colt's life and legacy, we will delve deeper into his early years, the development of his revolutionary revolver, its impact on warfare and society at large, as well as his lasting contributions to American industry. Through this exploration, we hope to shed light on one man's remarkable journey from a curious young boy to an iconic figure whose inventions forever changed the world.

***

The life of Samuel Colt, the renowned American inventor and industrialist, was greatly influenced by his childhood experiences. Born on July 19, 1814, in Hartford, Connecticut, Colt displayed remarkable curiosity and ingenuity from an early age. These formative years would lay the foundation for his future achievements and establish him as one of the most influential figures in firearm history.

Colt's fascination with mechanics was evident even during his childhood. Growing up on a farm, he spent countless hours tinkering with machinery and experimenting with various contraptions. His upbringing in a rural environment exposed him to the practical applications of tools and machines, sparking his imagination and fueling his passion for invention.

Furthermore, Colt's exposure to firearms at an early age played a crucial role in shaping his future endeavors. His father owned a successful textile mill where he produced woolen goods for military uniforms. This connection to the military world exposed young Samuel to firearms regularly. He observed how soldiers handled their weapons and became captivated by their power and functionality.

It was during these early encounters that Colt began envisioning ways to improve upon existing firearm designs. He noticed inefficiencies in manual reloading processes that hindered soldiers' ability to fire rapidly in combat situations. This observation would later inspire him to develop a revolutionary mechanism that allowed for rapid firing without manual reloading – an innovation that would change the course of warfare forever.

In addition to his mechanical interests, Colt's childhood experiences also instilled in him an entrepreneurial spirit. Growing up amidst the Industrial Revolution, he witnessed firsthand how innovative ideas could transform industries and create wealth. This exposure fueled his ambition to become not just an inventor but also a successful businessman.

Colt's father recognized his son's potential and encouraged him to pursue education beyond what was typically available at that time. As a result, Samuel attended various private schools where he gained knowledge in science, mathematics, and engineering. These academic pursuits further honed his technical skills and provided him with a solid foundation for his future endeavors.

However, it was not just his formal education that shaped Colt's future; it was also the lessons he learned from failures and setbacks. As a young boy, he faced financial hardships when his father's business faltered. Witnessing firsthand the consequences of economic instability drove Colt to develop a resilient and determined mindset – traits that would later prove invaluable in overcoming challenges throughout his entrepreneurial journey.

Samuel Colt's early life experiences played a pivotal role in shaping his future achievements. His fascination with mechanics, exposure to firearms, entrepreneurial spirit, and education all contributed to the development of an innovative mind and an unwavering determination. These qualities would be instrumental in propelling him towards inventing one of the most iconic firearms in history – the Colt revolver – forever cementing his legacy as a pioneering figure in both industry and technology.

***

In the world of firearms, few names carry as much weight and historical significance as Samuel Colt. Born on July 19, 1814, in Hartford, Connecticut, Colt would go on to become one of the most influential figures in firearm design and manufacturing. His inventions revolutionized the industry and forever changed the way people think about handguns. At the heart of his success was the creation of the first practical revolver.

Colt's fascination with guns began at a young age when he observed his father working with a flintlock pistol. This early exposure sparked his passion for firearms and set him on a path to innovate within this field. However, it wasn't until he traveled abroad that he gained inspiration for his groundbreaking invention.

During a voyage to India in 1830, Colt noticed that British officers carried single-shot pistols while their adversaries wielded multiple-shot weapons like muskets. This observation planted a seed in Colt's mind – an idea that would eventually lead to the birth of his iconic revolver.

Upon returning to America, Colt dedicated himself to perfecting a revolving cylinder mechanism that allowed for multiple shots without reloading. After years of experimentation and numerous prototypes, he finally patented his design in 1836. The following year marked the birth of Colt Firearms Company and the production of their first commercially successful model – the Colt Paterson.

The Paterson revolver featured a five- or six-shot cylinder that rotated automatically with each pull of the trigger. This innovative mechanism eliminated several limitations associated with single-shot pistols at that time. It offered quick reloads by simply swapping out pre-loaded cylinders instead of manually loading each chamber individually – an efficiency unparalleled until then.

Colt's invention gained considerable attention both domestically and internationally due to its reliability and ease of use compared to other contemporary firearms. Its impact was felt far beyond military applications; it provided civilians with a reliable and effective means of self-defense.

With the success of the Paterson, Colt continued to refine his designs, constantly pushing the boundaries of firearm technology. By introducing improvements such as interchangeable parts and mass production techniques, he not only revolutionized firearm manufacturing but also set the stage for industrialization in other sectors.

The Colt revolver gained widespread popularity during the Texas Revolution and later became synonymous with the American West. It played a significant role in shaping American history, from its use by lawmen like Wild Bill Hickok and Wyatt Earp to its adoption by military forces during conflicts such as the Mexican-American War.

Samuel Colt's contributions to firearms technology cannot be overstated. Through his relentless pursuit of innovation, he forever changed the landscape of handguns. His invention of the first practical revolver laid a foundation upon which future firearms advancements were built. Today, Colt Firearms Company continues to produce high-quality firearms that bear Samuel Colt's name – a testament to his enduring legacy in the world of weaponry.

***

When discussing the wild and untamed era of the American West, it is impossible to overlook the profound impact of Samuel Colt and his revolutionary invention, the Colt Revolver. With its unmatched reliability, speed, and firepower, this firearm played a pivotal role in transforming a lawless frontier into a more civilized society.

Prior to Colt's invention, firearms were often unreliable and slow to reload. This limited their effectiveness in confrontations with outlaws and hostile Native American tribes. However, Colt's revolver changed everything. Its innovative design allowed for multiple shots without reloading, giving its wielder a significant advantage in any encounter.

One of the most significant contributions of Colt's revolver was its ability to level the playing field. In an era where physical strength often determined one's survival in a land filled with danger and lawlessness, this firearm became an equalizer. It enabled individuals who were not physically imposing to defend themselves against larger adversaries. This newfound sense of empowerment had far-reaching effects on society as it encouraged individualism and self-reliance among settlers.

The Colt Revolver also played a crucial role in law enforcement during this tumultuous period. Lawmen such as Wyatt Earp and Bat Masterson relied heavily on these revolvers to maintain order in towns that were often overrun by criminals. The firepower provided by these weapons gave lawmen an advantage over outlaws who previously held sway over these territories.

Additionally, the presence of reliable firearms like Colt's revolver acted as a deterrent for potential criminals. The knowledge that citizens were armed with such potent weapons made them think twice before engaging in criminal activities. As a result, communities became safer overall as individuals began to respect each other's rights and property.

Furthermore, Colt's revolver had significant implications for Native American tribes during this time period. While many tribes initially possessed superior knowledge of warfare tactics compared to settlers, they were often outmatched when it came to firearms. The Colt Revolver allowed settlers to defend themselves against Native American attacks, ultimately leading to the conquest of their lands and the subjugation of many tribes.

The impact of Colt's revolver on the Wild West extended beyond its immediate effects on law enforcement and self-defense. It also played a role in shaping American culture and mythology. The image of a cowboy armed with a Colt Revolver became synonymous with the rugged individualism and frontier spirit that defined this era. This iconic firearm is forever ingrained in popular culture, serving as a symbol of both lawlessness and justice, depending on one's perspective.

Samuel Colt's invention had an indelible impact on the wild west. By providing settlers with a reliable, fast-firing weapon, his revolver transformed the frontier from an untamed wilderness into a more civilized society. It empowered individuals while simultaneously aiding law enforcement efforts. Moreover, it played a role in shaping American culture and remains an enduring symbol of this tumultuous period in history.

***

In the annals of firearms history, few names command as much respect and admiration as Samuel Colt. His innovative designs revolutionized the world of weaponry, forever changing the course of warfare and self-defense. Among his many groundbreaking creations, none stand out quite like the iconic Colt Peacemaker – a legendary six-shooter that has become synonymous with American frontier lore.

Born in 1814, Samuel Colt was destined to leave an indelible mark on firearms design. His relentless pursuit of perfection led him to create one of his most enduring masterpieces – the Colt Single Action Army revolver, commonly known as the Peacemaker. Introduced in 1873, this formidable handgun quickly became a symbol of power and authority throughout the Wild West.

The Colt Peacemaker boasted an array of features that set it apart from its contemporaries. Its single-action mechanism allowed for quick and accurate shooting by manually cocking the hammer before each round was fired. With its robust construction and simple design, this six-shooter proved to be highly reliable even under challenging conditions.

One notable aspect of the Peacemaker was its chambering for .45 Long Colt ammunition – a powerful cartridge that delivered exceptional stopping power at close range. This made it a favorite among lawmen, outlaws, and cowboys alike, who relied on its ability to neutralize threats swiftly and effectively.

Moreover, the aesthetic appeal of the Peacemaker contributed greatly to its iconic status. The revolver featured a distinctive barrel length ranging from four to seven and a half inches, with a sleek blue or nickel-plated finish accentuating its elegant lines. The gun's ivory or walnut grips added further elegance while ensuring a firm grasp during rapid shooting sequences.

The impact of this legendary firearm extended far beyond its practical applications; it transcended into popular culture through countless books and movies portraying life in the Old West. The Peacemaker's association with legendary figures such as Wyatt Earp and Wild Bill Hickok only heightened its allure, solidifying its status as a symbol of law and order in an untamed land.

While the Colt Peacemaker undoubtedly played a significant role in shaping the American West, its legacy goes beyond historical significance. Today, it is revered as a collector's item, fetching high prices at auctions and gun shows worldwide. Its timeless design and undeniable influence on subsequent revolver models cemented its place in the pantheon of firearms history.

The Colt Peacemaker stands out as one of Samuel Colt's crowning achievements. This legendary six-shooter represents not only a pinnacle of firearm engineering but also an embodiment of America's frontier spirit. Its enduring popularity and cultural impact serve as a testament to Samuel Colt's ingenuity, forever immortalizing him as one of history's greatest firearms designers.

***

Samuel Colt, an ingenious inventor and entrepreneur, forever changed the landscape of American industry with his revolutionary firearms. Through innovation, strategic marketing, and a commitment to quality, Colt Firearms experienced remarkable expansion and success during the 19th century, leaving an indelible mark on both the arms manufacturing sector and American history.

One of the key factors behind Colt Firearms' transformative impact was Samuel Colt's innovative design for a revolving cylinder pistol. This groundbreaking invention allowed for multiple shots without reloading, providing a significant advantage over traditional single-shot firearms. Recognizing the potential of his invention, Colt embarked on an ambitious journey to bring his vision to life.

In 1836, Samuel Colt founded the Patent Arms Manufacturing Company in Paterson, New Jersey. Despite initial setbacks and financial difficulties due to skepticism surrounding his new firearm design, Colt persevered. He secured military contracts through shrewd marketing strategies that showcased the efficiency and reliability of his revolvers. These contracts not only provided financial stability but also served as powerful endorsements that bolstered public trust in his firearms.

Colt's commitment to quality was another vital aspect that propelled his company's success. With meticulous attention to detail and rigorous testing protocols, he ensured that each revolver leaving the factory met stringent standards of performance and durability. This dedication earned him a reputation for producing some of the finest firearms in America.

As demand grew for their products both domestically and internationally, Colt Firearms expanded its manufacturing capabilities significantly. In 1847, it relocated its operations from Paterson to Hartford, Connecticut – a move that allowed for increased production capacity due to better access to resources such as skilled laborers and raw materials.

Furthermore, Samuel Colt embraced technological advancements within the industry by implementing assembly line production techniques as early as 1851. This innovative approach dramatically increased output while reducing costs – allowing him to offer competitive prices without compromising quality.

The impact of Colt Firearms on American industry extended beyond the firearms sector. The company's manufacturing practices, such as interchangeable parts and the assembly line, became influential examples for other industries seeking to streamline production processes and increase efficiency. This revolutionized manufacturing in America, paving the way for future advancements in various sectors.

Moreover, Colt's success contributed significantly to the economic growth of Hartford, Connecticut. The expansion of his manufacturing operations brought employment opportunities and attracted skilled workers to the area. As a result, Hartford developed into a thriving industrial hub during the 19th century.

Samuel Colt's relentless pursuit of innovation, strategic marketing efforts, commitment to quality, and adoption of efficient manufacturing practices led to Colt Firearms' remarkable expansion and success. His revolutionary revolver design not only transformed the arms industry but also influenced other sectors by setting new standards for production efficiency. Samuel Colt's legacy as an industrial pioneer remains an integral part of American history and serves as a testament to his enduring impact on American industry.

***

Samuel Colt, renowned as the inventor of the revolver, made significant contributions to various industries beyond his iconic firearm. While his revolver revolutionized firearms technology, Colt's innovative spirit extended far beyond this invention. Throughout his career, he patented numerous other inventions that had a lasting impact on fields such as engineering, manufacturing, and even naval warfare.

One of Colt's notable contributions was his design for underwater mines or torpedoes. In 1836, he patented a device called a "Submarine Battery," which was essentially an explosive-filled barrel that could be submerged in water and detonated remotely. This invention played a crucial role in naval warfare by providing an effective means to defend harbors and block enemy ships from entering.

As a result of this innovation, Colt received several contracts from the U.S. government to manufacture these torpedoes during times of conflict.

Colt also made strides in improving printing press technology with his invention of the "Colt Armory Press." This press utilized steam power to operate rather than relying solely on manual labor. By automating certain processes involved in printing newspapers and books, it significantly increased productivity and efficiency in the publishing industry. This innovation was widely adopted across printing presses worldwide and played a pivotal role in modernizing the field.

Another area where Colt left his mark was underwater cable-laying technology. In 1854, he patented a system for laying telegraph cables on the ocean floor—a process that had previously been fraught with difficulties and limitations. His design incorporated specialized machinery that allowed for efficient cable deployment while minimizing damage or entanglement during installation—a significant breakthrough at the time. This innovation facilitated faster communication across vast distances through submarine telegraph cables—a vital development for global communication networks.

Colt also ventured into firearm manufacturing techniques by patenting improvements related to interchangeable parts production methods—an area where he excelled due to his engineering background. His innovative approach to mass production allowed for the creation of firearms with interchangeable components, making repairs and replacements significantly easier. This breakthrough not only enhanced the reliability and longevity of firearms but also revolutionized manufacturing processes across various industries.

Furthermore, Colt's contributions extended to other inventions, such as a floating battery system for river navigation, a hydraulic press for compressing cotton, a rotary engine design, and even a reaping machine. These inventions showcased his versatility as an inventor and his eagerness to explore diverse fields beyond firearms.

Samuel Colt's impact on innovation extended well beyond the realm of revolvers. His patents and inventions in underwater mines, printing press technology, cable laying techniques, interchangeable parts production methods, and various other fields demonstrate his pioneering spirit and creativity. Colt's legacy as an inventor remains significant today; his contributions have shaped industries worldwide and continue to influence technological advancements in numerous domains.

***

While Samuel Colt is widely regarded as a pioneer in the field of firearms manufacturing, his career was not without challenges and controversies. As with many influential figures, Colt faced criticism and scrutiny throughout his life. This subtopic aims to delve into some of the key criticisms faced by Samuel Colt, shedding light on the complexities of his legacy.

One of the most prevalent criticisms directed at Colt was his aggressive patenting practices. He held numerous patents for firearm designs and manufacturing processes, which allowed him to establish a monopoly over the industry. This dominance led to accusations of stifling innovation and hindering competition. Critics argued that Colt's patents prevented other inventors from improving upon or developing alternative firearm technologies, ultimately limiting progress within the industry.

Furthermore, some critics accused Samuel Colt of exploiting laborers in his factories. During the mid-19th century, when industrialization was rapidly transforming American society, labor conditions were often harsh, and workers' rights were frequently neglected. While it is true that Colt's factories provided employment opportunities for thousands of individuals, reports suggest that he paid low wages and subjected workers to long hours under demanding conditions.

These allegations tarnished his reputation as an industrialist and philanthropist.

Another controversy surrounding Samuel Colt relates to his involvement in international arms sales. Despite being an American inventor and manufacturer, he sought to expand his market globally by supplying firearms to foreign governments during times of conflict. This raised ethical concerns about profiting from warfare and fueling violence overseas. Critics argued that rather than contributing to peace or security, Colt's actions perpetuated a cycle of violence by arming nations engaged in conflicts.

Additionally, there were accusations that Samuel Colt exaggerated or fabricated aspects of his personal history for self-promotion purposes. Some claimed that he embellished stories about his early life experiences or achievements in order to enhance his image as an entrepreneur and inventor. While it is common for successful individuals to embellish their stories, these allegations cast doubt on the authenticity of Colt's narrative and the accuracy of his self-portrayal.

Samuel Colt faced several challenges and controversies throughout his career, which shaped public perception of him. His aggressive patenting practices, exploitation of laborers, involvement in international arms sales, and alleged embellishments of personal history were key criticisms directed at him. While Colt undoubtedly made significant contributions to the firearms industry and played a crucial role in American history, it is essential to examine these controversies to gain a more comprehensive understanding of his complex legacy.

***

Samuel Colt, an American inventor and industrialist, revolutionized the world of firearms with his innovative designs and manufacturing techniques. His contributions to gun manufacturing have left an indelible mark on the industry, shaping it into what it is today. This article explores the lasting legacy of Samuel Colt and how his inventions continue to influence gun manufacturing.

One of Colt's most significant contributions was his invention of the revolver. Prior to Colt's design, guns were single-shot muskets or rifles that required reloading after each shot. The revolver, with its rotating cylinder capable of holding multiple rounds, allowed for rapid firing without the need for constant reloading. This breakthrough in firearm technology forever changed combat tactics and self-defense strategies.

Colt's innovative manufacturing methods also played a crucial role in his legacy. He introduced assembly line production techniques to firearm manufacturing, streamlining the process and increasing efficiency. By standardizing parts and using interchangeable components, he made mass production feasible and affordable. This approach not only transformed gun manufacturing but also influenced other industries worldwide.

The impact of Colt's inventions extended beyond America's borders. His revolvers gained global recognition for their reliability and superior performance during conflicts such as the Mexican-American War and the American Civil War. These firearms became synonymous with power, precision, and reliability – qualities sought after by military forces around the world.

Furthermore, Samuel Colt's business acumen contributed significantly to his legacy. He established a successful company that bore his name – Colt's Patent Fire-Arms Manufacturing Company – which became one of America's most iconic firearms manufacturers. Under his guidance, the company grew rapidly and gained a reputation for quality craftsmanship.

Colt also understood marketing strategies before they became mainstream practices. He employed various advertising techniques to promote his products effectively – from engaging demonstrations showcasing their capabilities to targeted advertisements in newspapers and magazines across America.

Beyond his immediate contributions, Samuel Colt's legacy continues to shape the gun manufacturing industry. His commitment to quality and innovation set a high standard that subsequent generations of firearm manufacturers strive to achieve. The principles he established – interchangeable parts, efficient production methods, and reliable designs – are still fundamental in modern gun manufacturing.

Moreover, Colt's impact extends beyond firearms themselves. His inventions influenced the development of other mechanical devices and machinery, as well as manufacturing techniques used in various industries today. From automobiles to appliances, the concept of interchangeable parts owes its origins to Samuel Colt's groundbreaking work.

Samuel Colt's lasting influence on gun manufacturing is undeniable. His revolutionary designs, innovative manufacturing methods, and business acumen have left an indelible mark on the industry. The legacy he created continues to shape firearm production worldwide and extends beyond guns themselves. Samuel Colt will forever be remembered as a legendary figure whose contributions continue to be celebrated by both historians and firearm enthusiasts alike.

***

Throughout history, certain individuals have left an indelible mark on their respective fields, forever changing the course of human progress. Samuel Colt, a visionary inventor and entrepreneur, stands among these remarkable figures. Through his pioneering work in firearm design and manufacturing, Colt revolutionized the industry and left a lasting legacy that continues to influence the world today. Samuel Colt's contributions to firearms technology were nothing short of extraordinary.

His most significant innovation was the creation of the revolving cylinder pistol, commonly known as the revolver. This groundbreaking design allowed multiple shots to be fired without having to manually reload after each shot—a game-changer in terms of both efficiency and firepower. The revolver quickly gained popularity among law enforcement agencies, military forces, and civilians alike due to its reliability, ease of use, and versatility.

Colt's genius not only lay in his ability to develop innovative designs but also in his relentless pursuit of perfection. He continually refined his inventions through rigorous testing and modifications until he achieved optimal performance. This unwavering commitment to excellence set a new standard for firearm manufacturing—a tradition that has been carried forward by countless gun manufacturers over the years. Beyond his technical achievements, Samuel Colt was also an astute businessman who understood the importance of marketing and branding.

He recognized early on that successful promotion was key to establishing a foothold in any industry. Colt leveraged various advertising strategies, such as endorsements by prominent figures and demonstrations showcasing his weapons' capabilities at exhibitions across America and Europe. These efforts not only generated widespread awareness but also solidified Colt's position as an industry leader. Colt's impact on society extended far beyond just firearms; he played a pivotal role in shaping American history itself.

During the mid-19th century, when America was expanding westward with increasing speed, settlers needed reliable weapons for protection against hostile Native American tribes as well as outlaws. Colt's revolvers became the weapon of choice for pioneers and lawmen alike, providing them with a sense of security and confidence in the face of adversity. Furthermore, Colt's inventions played a significant role in the outcome of many historical conflicts.

The Colt Walker revolver, introduced during the Mexican-American War and later utilized during the American Civil War, gave soldiers an unprecedented advantage on the battlefield. Its sheer power and rapid-fire capability turned the tide in numerous engagements, solidifying its place as one of history's most influential weapons. As we reflect on Samuel Colt's enduring contributions to firearms technology, it becomes evident that his impact is still felt today.

Revolvers remain popular among shooters worldwide due to their reliability and distinctive design elements that have become synonymous with firearms themselves. Additionally, Colt's legacy lives on through his eponymous company, which continues to produce high-quality firearms that honor his original vision. Samuel Colt's contributions to the world of firearms cannot be overstated. His revolutionary designs continue to shape both the industry and society as a whole.

Through his unwavering commitment to innovation, relentless pursuit of perfection, and astute marketing strategies, Colt forever cemented himself as one of history's greatest inventors.

Louis-Nicolas Flobert
1819-1894

In the annals of history, there are certain individuals whose ingenuity and creativity have forever altered the course of human progress. One such luminary is Louis-Nicolas Flobert, a brilliant French inventor whose pioneering work in firearms revolutionized the way we perceive and utilize these powerful tools. From humble beginnings to worldwide recognition, Flobert's life and legacy remain an intriguing tale that deserves to be explored.

Born on October 14, 1819, in Paris, France, Flobert displayed an innate curiosity from a young age. Growing up during a period marked by rapid industrialization and technological advancements, he was exposed to an array of scientific disciplines that would shape his future endeavors. However, it was his passion for firearms that would ultimately define his legacy.

At a time when traditional muzzle-loading guns were prevalent, Flobert sought to create a safer and more efficient alternative. His breakthrough came in 1845 when he invented the first commercially successful rimfire metallic cartridge – a cartridge with priming compound placed around its rim rather than within its center. This groundbreaking innovation allowed for easier reloading and more reliable ignition compared to existing firearm mechanisms.

Flobert's creation paved the way for numerous advancements in firearms technology that followed. His design laid the foundation for later developments such as centerfire cartridges – where the priming compound is located at the center – which became widely used in modern ammunition. Moreover, his invention was instrumental in transitioning firearms from single-shot weapons to repeaters capable of firing multiple rounds without reloading.

Beyond his technical achievements, Flobert's revolutionary invention had profound implications for various fields beyond military applications. Target shooting became more accessible as people could now practice their marksmanship skills without cumbersome reloading procedures or expensive equipment. This newfound accessibility led to increased interest in shooting sports across Europe and eventually worldwide, contributing to the growth of shooting clubs and competitions.

Despite the transformative impact of his invention, Flobert himself remained a relatively obscure figure during his lifetime. However, his legacy endured through the continued development and refinement of his cartridge design by subsequent inventors. Today, Flobert cartridges are still used in various firearms for a range of purposes, including training, sport shooting, and pest control.

As we delve deeper into the life and legacy of Louis-Nicolas Flobert, we uncover not only an inventor but also a visionary who forever changed the course of firearms history. Through his relentless pursuit of innovation and dedication to making firearms safer and more accessible, Flobert's genius persists in every modern firearm that bears a resemblance to his original concept.

In this exploration of Louis-Nicolas Flobert's life and legacy, we will trace the trajectory of his career as an inventor while delving into the societal impact brought about by his groundbreaking invention. From examining the challenges, he faced during a time marked by political upheaval to unraveling the far-reaching consequences that continue to shape our world today, this captivating journey aims to shed light on one man's remarkable contributions to human progress.

***

Louis-Nicolas Flobert, the visionary inventor who revolutionized firearms, had a remarkable journey that led him to create one of the most significant advancements in the history of weaponry. Born on April 2, 1819, in Paris, France, Flobert grew up in a time when technological innovations were flourishing, paving the way for his own pioneering contributions. Flobert's passion for invention can be traced back to his childhood.

From a young age, he showed an innate curiosity and mechanical aptitude that set him apart from his peers. As the son of a gunsmith, he was immersed in an environment that fueled his fascination with firearms. His father's workshop became his playground—a space where he could observe and experiment with various weapons. However, it was not only the physical aspects of guns that captivated Flobert; it was their inner workings as well.

He was particularly interested in understanding how gunpowder ignited and propelled bullets forward. This fascination with explosive power would become a driving force behind his later achievements. Flobert's early education played a crucial role in shaping his intellectual development. Enrolled at Collège Stanislas in Paris—an institution renowned for its emphasis on science and mathematics—he thrived under the guidance of talented teachers who recognized and nurtured his potential.

It was during this period that he gained a solid foundation in scientific principles and acquired problem-solving skills that would prove invaluable throughout his career. Furthermore, Flobert found inspiration outside formal education through extensive reading. He devoured books on mechanics, physics, chemistry, and engineering—expanding his knowledge beyond what traditional schooling could offer him. These self-guided studies allowed him to explore diverse fields of science and develop a multidisciplinary approach to problem-solving.

As Flobert entered adulthood, his natural inclination towards innovation led him to experiment with his own ideas. He began by modifying existing firearms, striving to improve their accuracy and reliability. These early attempts paved the way for his groundbreaking invention—the Flobert rifle. In 1845, at the age of 26, Flobert introduced the world to his revolutionary firearm—a small-caliber breech-loading rifle that would become known as the Flobert rifle.

The weapon featured an innovative, self-contained metallic cartridge, which contained both the bullet and gunpowder in a single unit. This design eliminated the need for loading individual components separately, significantly increasing efficiency and ease of use. The Flobert rifle's impact on firearms technology cannot be overstated. It laid the foundation for modern-day ammunition systems and influenced subsequent developments in firearms design.

Its success propelled Louis-Nicolas Flobert into prominence within the scientific community and secured his place in history as one of the most influential inventors in the field of weaponry. Louis-Nicolas Flobert's journey from a curious child to a renowned inventor was marked by a combination of innate talent, an exceptional education, and an insatiable thirst for knowledge.

***

In the world of firearms, few inventions have had as significant an impact as the Flobert cartridge. This revolutionary innovation, created by the brilliant mind of Louis-Nicolas Flobert, forever changed the way firearms were designed and used. By delving into the life and legacy of Louis-Nicolas Flobert, we can gain a deeper understanding of this game-changing invention and its lasting influence on the world.

Born in Paris in 1819, Louis-Nicolas Flobert displayed an early interest in mechanics and engineering. As he grew older, his passion for firearms led him to experiment with various designs and mechanisms. However, it was not until 1845 that he would unveil his most groundbreaking creation - the Flobert cartridge.

The Flobert cartridge was a self-contained unit comprising a bullet, propellant charge, and primer, all housed within a metallic casing. This ingenious design allowed for easy loading and increased safety while firing. Prior to this invention, shooters had to manually measure out powder charges and load them into separate chambers or barrels.

What truly set the Flobert cartridge apart from its predecessors was its small caliber size. While previous cartridges were primarily designed for larger firearms such as rifles or pistols, Flobert focused on creating a small-caliber round suitable for indoor shooting or training purposes. The reduced power of these cartridges made them ideal for target practice without causing excessive noise or recoil.

Flobert's invention quickly gained popularity among shooting enthusiasts across Europe. Target shooting became accessible to a wider audience due to the affordability and ease-of-use offered by these small-caliber cartridges. In addition to recreational use, they also found practical applications in varmint control and pest management.

The impact of the Flobert cartridge extended beyond just its immediate success; it laid the foundation for further advancements in firearms technology. Flobert's design served as a precursor to the rimfire cartridge, which became widely adopted in the late 19th century. This development marked a significant leap forward in ammunition design and facilitated the creation of more compact and versatile firearms.

Louis-Nicolas Flobert's groundbreaking invention not only revolutionized firearms but also left an indelible mark on the history of shooting sports. The accessibility and convenience provided by the Flobert cartridge allowed people from all walks of life to engage in target shooting, fostering a love for marksmanship that endures to this day.

The birth of the Flobert cartridge was a pivotal moment in firearms history. Louis-Nicolas Flobert's ingenuity and determination led to the creation of an innovation that forever changed how firearms were used and enjoyed. By exploring his life and legacy, we can appreciate the immense impact this game-changing invention had on shooting sports and ammunition development.

***

The invention of Louis-Nicolas Flobert, a French gunsmith and inventor, had a profound impact on the world of firearms. Flobert's creation, known as the Flobert Parlor Pistol, not only revolutionized gun design but also laid the foundation for future advancements in firearms technology. This article will explore how Flobert's invention transformed the world of guns, revolutionizing both their design and their usage.

Before Flobert's innovation, firearms were primarily used for military purposes and hunting large game. The introduction of the Flobert Parlor Pistol marked a significant shift in this paradigm. Designed as a small-caliber firearm suitable for indoor shooting, this invention opened up new possibilities for recreational shooting and target practice. The pistol was specifically tailored to be used in confined spaces such as parlors or shooting galleries, allowing enthusiasts to practice their marksmanship skills without venturing outdoors.

Flobert achieved this by designing a compact firearm that utilized low-powered ammunition. The pistol fired small lead bullets propelled by rimfire cartridges - an ingenious mechanism that allowed for indoor use while minimizing noise and recoil. This breakthrough made recreational shooting accessible to a wider audience, including women and people with limited access to outdoor shooting ranges.

Furthermore, the introduction of the Flobert Parlor Pistol paved the way for subsequent innovations in firearms technology. It acted as a precursor to modern-day rimfire cartridges commonly used in handguns today. These cartridges feature self-contained priming compounds located within their rims - an idea initially conceived by Flobert himself.

The impact of this revolutionary design did not stop there; it also influenced future developments in firearm safety mechanisms. As demand grew for more powerful handguns capable of firing larger calibers outdoors, engineers looked to improve upon Flobert's initial concept while ensuring user safety remained paramount. This led to the development of new firing mechanisms, such as centerfire cartridges, which utilized a centrally located primer for increased reliability and reduced chances of accidental discharge.

Flobert's invention not only transformed the design and usage of firearms but also played a significant role in shaping the cultural perception surrounding guns. The Flobert Parlor Pistol introduced shooting as a recreational activity, distancing it from its previous associations solely with warfare or hunting. This shift allowed firearms to be seen as tools for sport and entertainment rather than just instruments of violence.

Louis-Nicolas Flobert's invention of the Flobert Parlor Pistol revolutionized the world of firearms in several ways. It opened up new avenues for recreational shooting and target practice by introducing a compact, low-powered firearm suitable for indoor use. Additionally, his innovative design laid the foundation for future advancements in firearms technology, influencing the development of modern rimfire cartridges and safety mechanisms.

By transforming the cultural perception surrounding guns, Flobert's invention paved the way for firearms to be viewed as tools for sport and entertainment rather than solely instruments of war or hunting.

***

Louis-Nicolas Flobert, a renowned French inventor, is widely celebrated for his groundbreaking contributions to the field of firearms. However, like many pioneers, he faced several challenges and controversies throughout his lifetime. This subtopic delves into the criticisms that were directed at Flobert during his career, shedding light on the obstacles he had to overcome.

One of the primary criticisms aimed at Flobert was his invention's potential for misuse. The introduction of the first practical breech-loading firearm, known as the "Flobert gun," sparked concerns among some individuals who feared that this innovation would facilitate criminal activities. Critics argued that these easily concealed firearms could be used by wrongdoers without detection or traceability, thereby endangering public safety.

Despite these concerns, Flobert remained steadfast in his belief that responsible usage and regulation could mitigate any potential risks associated with his invention.

Another criticism levied against Flobert was rooted in concerns regarding accuracy and reliability. Traditional muzzle-loading firearms have been relied upon for centuries due to their proven track record in accuracy. Skeptics doubted whether Flobert's breech-loading mechanism could match or surpass their precision. These critics argued that precision shooting required a stable platform offered by muzzle-loaders and expressed doubts about whether a breech-loading design could deliver consistent results.

Nevertheless, Flobert's perseverance paid off as subsequent advancements built upon his work further improved accuracy and reliability.

Furthermore, some critics accused Flobert of merely modifying existing designs rather than genuinely revolutionizing firearms technology. They contended that while his inventions were undoubtedly innovative in their own right, they fell short of constituting a true revolution in firearm design and functionality. Detractors argued that other inventors had made more substantial contributions to the field through their radical departures from conventional firearms.

However, Flobert's legacy demonstrates that even incremental improvements can have a lasting impact on an industry.

Flobert also faced challenges related to the commercial viability of his inventions. Critics questioned the market demand for his breech-loading firearms, believing that traditional muzzle-loaders would continue to dominate the market. The initial high cost associated with manufacturing these new firearms posed an additional obstacle, as it limited their accessibility to a broader customer base. However, Flobert's perseverance and entrepreneurial spirit eventually led to the widespread adoption of his designs, proving that there was indeed a market for his innovative firearms.

Louis-Nicolas Flobert encountered various criticisms and challenges throughout his lifetime as he revolutionized firearm technology. Concerns regarding potential misuse, doubts about accuracy and reliability, accusations of not truly revolutionizing the field, and skepticism over market demand were among the key controversies he faced. Despite these obstacles, Flobert's groundbreaking contributions paved the way for future advancements in firearms design and solidified his position as one of history's notable inventors in this field.

***

While Louis-Nicolas Flobert is primarily known for his revolutionary contributions to firearms, his innovations extended far beyond the realm of weaponry. Flobert's inventive mind and dedication to pushing boundaries resulted in applications that went beyond traditional firearms and impacted various fields, including mechanics, engineering, and even sports. This subtopic will delve into some of these diverse applications and shed light on the multifaceted legacy left by this remarkable inventor.

One notable application of Flobert's innovations lies in the field of mechanics. His pioneering work with cartridges, specifically the development of rimfire ammunition, revolutionized not only firearms but also other mechanical devices. Rimfire ammunition involves a small primer compound located within the rim of a cartridge case, which ignites upon striking. This ingenious design opened doors for new possibilities in mechanisms such as ignition systems for engines or even time-delayed devices.

Flobert's mechanical genius also found its way into engineering applications. The precision and reliability he sought in his firearm designs led to advancements in machining techniques and tolerances. His pursuit of perfection pushed engineers to develop more precise tools and machinery that were essential not only for firearm production but also for general manufacturing processes across industries.

Additionally, Flobert's inventions had an unexpected impact on recreational activities. The introduction of air guns, based on his principles, allowed people to engage in shooting sports without relying on traditional firearms or gunpowder. These air guns utilized compressed air or gas to propel projectiles with accuracy and precision, similar to their powder-based counterparts. This breakthrough made shooting sports more accessible while promoting safety and reducing noise pollution.

Moreover, Flobert's innovations inspired further advancements in marksmanship training techniques. The introduction of indoor shooting galleries using low-power cartridges derived from his designs provided individuals with a controlled environment to practice their aim without the need for expansive outdoor ranges or expensive live ammunition. This innovation democratized marksmanship training and allowed enthusiasts to hone their skills in a more convenient and cost-effective manner.

Beyond mechanics, engineering, and recreational applications, Flobert's inventions had an enduring impact on the development of firearms as art forms. His pursuit of elegance and aesthetics in firearm design elevated them from mere tools to works of art. The balance between functionality and beauty that Flobert achieved inspired subsequent generations of gunsmiths to create firearms that were not only highly functional but also visually stunning.

The influence of Louis-Nicolas Flobert's innovations extends far beyond firearms. His contributions to mechanics, engineering, recreational activities, marksmanship training, and even artistic firearm design have left an indelible mark on various fields. By exploring the diverse applications of Flobert's innovations, we gain a deeper appreciation for his inventive spirit and the wide-ranging impact he had on society.

***

The life and legacy of Louis-Nicolas Flobert, the innovative inventor who revolutionized firearms, have left an indelible mark on modern firearms technology. Flobert's groundbreaking inventions and designs paved the way for numerous advancements that continue to shape the industry today.

One of Flobert's most significant contributions was the invention of the rimfire cartridge. Prior to his innovation, firearms primarily relied on percussion caps or flintlocks for ignition. Flobert's rimfire cartridge, which contained a priming compound within its rim, allowed for a more reliable and efficient ignition system. This breakthrough laid the foundation for future developments in ammunition design and significantly influenced subsequent firearm technology.

Flobert's invention also played a pivotal role in shaping the development of small-caliber firearms. His creation allowed for smaller and lighter weapons to be produced, making them more accessible and practical for various purposes such as hunting or self-defense. The introduction of small-caliber firearms marked a turning point in firearm design and greatly influenced subsequent innovations.

Furthermore, Flobert's work led to advancements in firearm manufacturing processes. His designs incorporated mass production techniques that streamlined production and reduced costs significantly. By introducing interchangeable parts into firearms manufacturing, he revolutionized the industry by enabling easier repairs and modifications.

Flobert's legacy extends beyond his technical innovations; he also had a profound impact on firearm safety measures. Recognizing that accidents could occur due to mishandling or unauthorized use of firearms, he developed safety mechanisms such as trigger guards and safety locks. These features were essential steps toward ensuring responsible firearm usage and minimizing accidental injuries.

Additionally, Flobert's inventive spirit extended beyond just firearms themselves; he also contributed to improvements in shooting ranges' design. He developed innovative target systems that allowed shooters to practice their aim more effectively while maintaining safety standards. These target systems were widely adopted and laid the groundwork for modern shooting ranges, providing enthusiasts with a controlled environment to hone their skills.

Flobert's contributions to firearms technology have had a lasting impact on the industry. His inventions and designs continue to influence modern firearm manufacturing processes, ammunition development, and safety measures. The rimfire cartridge he invented remains a popular choice for small-caliber firearms due to its reliability and versatility.

Moreover, Flobert's emphasis on mass production techniques has become standard practice in the industry, allowing for more efficient manufacturing processes that can meet growing demand. The safety mechanisms he introduced have become integral components of firearm design, ensuring responsible usage and reducing accidents.

Louis-Nicolas Flobert's legacy in innovation has left an indelible mark on modern firearms technology. His groundbreaking inventions, such as the rimfire cartridge, advancements in small-caliber firearms, improvements in manufacturing processes, and emphasis on firearm safety, have significantly shaped the industry. Flobert's contributions continue to influence the design and functionality of firearms today, showcasing his enduring impact as an inventor who revolutionized the field of firearms technology.

***

In the realm of firearms, certain names have become synonymous with innovation and revolution. Samuel Colt, John Browning, and Hiram Maxim are just a few examples of inventors whose contributions have shaped the modern world. However, there is one name that often goes unnoticed in the annals of history: Louis-Nicolas Flobert. Despite his significant impact on firearms development, Flobert remains a relatively obscure figure.

This subtopic aims to shed light on why his name continues to be lesser known in history books. One possible reason for Flobert's relative obscurity is the nature of his invention. While many inventors focused on creating new types of firearms or improving existing designs, Flobert's contribution was more subtle but no less groundbreaking. He is credited with inventing the first practical rimfire cartridge in 1845, which laid the foundation for modern ammunition technology.

However, compared to the dramatic advancements made by other firearm pioneers during that era, such as the introduction of repeating rifles or machine guns, Flobert's achievement may have appeared less significant at first glance. Another factor contributing to Flobert's anonymity could be attributed to his personal circumstances and lack of marketing prowess. Unlike some other inventors who actively promoted their creations or formed successful business ventures around them, Flobert lived a relatively quiet life away from public attention.

He primarily focused on perfecting his designs rather than seeking fame or financial gain from them. Consequently, this lack of self-promotion may have led to his inventions being overshadowed by those who were more adept at capturing public interest. Furthermore, historical biases can also play a role in shaping which figures receive recognition and which remain forgotten by mainstream narratives. During Flobert's time and for much of history thereafter, firearms were predominantly associated with warfare and the military.

This focus on the martial aspects of firearms innovation may have led to a disregard for Flobert's contributions, as his inventions were primarily intended for recreational shooting and target practice. The perceived lack of direct relevance to military applications might have relegated Flobert's name to the margins of historical discourse. Lastly, it is worth considering that Flobert's legacy may have been overshadowed by subsequent advancements in firearms technology.

While his invention of the rimfire cartridge was undoubtedly groundbreaking at the time, it was soon surpassed by other developments such as centerfire cartridges and self-contained metallic cartridges. These subsequent innovations captured the public imagination and became more closely associated with significant leaps forward in firearm technology, leaving Flobert's contribution somewhat overlooked. Louis-Nicolas Flobert's relative obscurity in history books can be attributed to several factors.

The nature of his invention, his personal circumstances, historical biases, and subsequent technological advancements all contribute to this oversight. However, it is essential to recognize his invaluable contribution as the inventor of the practical rimfire cartridge – a technological breakthrough that revolutionized ammunition development and paved the way for future innovations in firearms design.

***

Louis-Nicolas Flobert, the French inventor who revolutionized firearms, left an indelible mark on future generations through his pioneering work. His inventions and innovations not only transformed the world of firearms during his time but also set the stage for the development of even more advanced and sophisticated weapons in later years. This subtopic aims to explore how Flobert's legacy influenced subsequent inventors and their contributions to the evolution of firearms.

One of the primary ways in which Flobert's work shaped future generations was through his introduction of self-contained cartridges. Before Flobert, firearms were loaded with separate powder, projectile, and primer components. His invention of a cartridge that integrated these elements into a single unit made reloading faster, more efficient, and safer. This concept became a foundation for future inventors who sought to improve upon it.

Building upon Flobert's innovation, other inventors expanded on the concept of self-contained cartridges to create even more advanced systems. Notably, Gustave Young took this idea further by introducing metallic cartridges that featured a brass casing holding both powder and primer. Young's advancements paved the way for modern ammunition as we know it today.

Flobert's influence extended beyond cartridge design. His invention of the first practical rimfire cartridge ignited new possibilities for firearm developments. The rimfire technology allowed for smaller caliber weapons that were cheaper and easier to manufacture than their predecessors. As a result, inventors like Samuel Colt capitalized on this advancement by creating more compact revolvers that gained widespread popularity across various industries.

Furthermore, Flobert played an instrumental role in shaping later inventors' understanding of ballistics – specifically bullet trajectory and accuracy. His experiments with different bullet designs led him to develop bullets with increased stability during flight, minimizing deviations from intended paths. This knowledge became invaluable to subsequent inventors who sought to refine and improve bullet performance.

Flobert's ideas also influenced advancements in firearm mechanisms. His introduction of the percussion cap, a small explosive charge that ignited the main propellant charge, set the stage for later developments such as the rimfire and centerfire ignition systems. These innovations allowed for faster and more reliable firing mechanisms, ultimately enhancing the overall performance of firearms.

The impact of Flobert's work on future generations is evident in numerous inventors' contributions to firearms evolution. From cartridge design improvements to advancements in ballistics and firing mechanisms, his pioneering inventions laid the groundwork for further innovation. As a testament to his influence, many of these subsequent inventors built upon Flobert's ideas and expanded them into entirely new realms.

Louis-Nicolas Flobert's life and legacy had a profound influence on future generations of firearm inventors. Through his revolutionary inventions like self-contained cartridges, advancements in ballistics, and innovations in firing mechanisms, he provided a solid foundation for later inventors to build upon. Flobert's contributions shaped the trajectory of firearms evolution by paving the way for even more significant developments that continue to impact our world today.

***

Louis-Nicolas Flobert, a name that may not be as widely recognized as some other inventors, but his contributions to the world of firearms cannot be understated. Through his innovative designs and relentless pursuit of perfection, Flobert revolutionized the way we perceive and use firearms. His visionary ideas paved the way for many advancements in firearm technology, leaving an indelible mark on history.

One cannot discuss the life and legacy of Louis-Nicolas Flobert without mentioning his most notable invention - the rimfire cartridge. This groundbreaking creation transformed firearms forever, introducing a reliable and efficient ignition system that would go on to become a staple in firearm design.

Flobert's ingenuity in developing this cartridge laid the foundation for countless innovations that followed.

Flobert's dedication to experimentation was unparalleled. He tirelessly explored new concepts, always striving to improve upon existing designs. His tireless efforts resulted in numerous inventions and modifications that significantly enhanced firearm functionality and safety. By pushing boundaries and challenging conventions, Flobert paved the way for future inventors to build upon his work.

Furthermore, Flobert's commitment to precision engineering set him apart from his contemporaries. He understood that accuracy was paramount when it came to firearms, leading him to develop mechanisms that ensured consistent performance with each shot fired. His attention to detail and insistence on perfection created a standard that remains central in modern firearm manufacturing.

Beyond his technical accomplishments, Louis-Nicolas Flobert also left behind a lasting legacy as an entrepreneur and businessman. He recognized not only the importance of innovation but also its commercial potential. By establishing successful manufacturing facilities, he brought his inventions into production at scale while creating employment opportunities for many.

Flobert's impact extended far beyond Europe; it reverberated across continents as his inventions spread worldwide. His rimfire cartridge became a staple in firearms of all types, from small pocket pistols to large-caliber rifles. Today, countless shooters and enthusiasts owe their enjoyment and safety to the groundbreaking contributions of Louis-Nicolas Flobert.

As we celebrate the enduring contributions of this visionary inventor, it is essential to recognize that his achievements were not without challenges. Flobert faced setbacks and obstacles throughout his career, but he never allowed them to deter him from pursuing his passion for firearms innovation. His resilience serves as an inspiration to all who strive for greatness despite adversity.

Louis-Nicolas Flobert's life and legacy stand as a testament to the transformative power of human ingenuity. Through his groundbreaking inventions, unwavering dedication, and entrepreneurial spirit, he revolutionized firearms forever. The enduring impact of his contributions is felt every time a firearm is fired or admired for its precision engineering. As we look back on the remarkable life of Louis-Nicolas Flobert, let us celebrate his visionary spirit and honor the legacy he left behind - one that continues to shape our world today.

Benjamin Tyler Henry
1821-1898

In the annals of American history, there are individuals whose contributions have forever altered the course of their respective fields. Benjamin Tyler Henry, a remarkable inventor and gunsmith of the 19th century, stands as one such luminary figure. Renowned for his revolutionary firearms, particularly the iconic Henry rifle, he left an indelible mark on both military strategy and civilian arms manufacturing.

This subtopic aims to explore the historical significance of Benjamin Tyler Henry and his groundbreaking firearms that transformed warfare and propelled technological advancements.

Born in 1821 in Claremont, New Hampshire, Benjamin Tyler Henry displayed an early aptitude for mechanical engineering. As a young man, he honed his skills working at various factories before joining the ranks of Robbins & Lawrence in Windsor, Vermont – a renowned firearms manufacturer at the time. It was here that Henry's ingenuity took flight as he revolutionized firearm design.

One cannot overstate the impact that Henry's most famous creation, the lever-action repeating rifle known as "the Henry," had on American history. Introduced during the Civil War era in 1860, this firearm represented a profound departure from traditional muzzle-loading rifles prevalent at that time. Its innovative design allowed for rapid-fire capability, with its tubular magazine holding up to sixteen metallic cartridges – a feat unheard of until then.

The strategic implications of such a firearm were immense. With its unparalleled firing speed and increased ammunition capacity compared to contemporary weapons like muskets or single-shot rifles, it provided soldiers with unmatched firepower on the battlefield. This transformational aspect played a pivotal role during critical engagements such as Gettysburg or Vicksburg when Union troops armed with Henry rifles achieved decisive victories over Confederate forces.

Beyond its military significance during wartime conflicts like the Civil War or Indian Wars that followed shortly after, Benjamin Tyler Henry's rifle also exerted profound influence on civilian arms manufacturing. The Henry rifle became synonymous with reliability, durability, and accuracy. Its popularity spread far and wide, making it a coveted weapon among frontiersmen, law enforcement agencies, and hunters alike.

Henry's achievements were not limited to firearms alone. He also invented the metallic cartridge, which remains an integral component of modern ammunition to this day. Before his innovation, firearms relied on loose powder and separate projectiles – a cumbersome and time-consuming process. By creating a self-contained cartridge with a primer, propellant, and bullet encased in one unit, Henry simplified the loading process and paved the way for further advancements in firearm technology.

Benjamin Tyler Henry's contributions to the world of firearms were nothing short of revolutionary. His invention of the lever-action repeating rifle – the Henry – forever changed military tactics during the Civil War while simultaneously revolutionizing civilian arms manufacturing. The strategic advantages it provided on the battlefield coupled with its exceptional reliability cemented its historical significance. Moreover, his creation of metallic cartridges advanced ammunition technology by leaps and bounds.

Benjamin Tyler Henry's legacy endures as an emblem of innovation that continues to shape our understanding of firearms today.

***

Benjamin Tyler Henry, the renowned firearms inventor and entrepreneur, is a figure of enduring fascination in American history. While his name is synonymous with the iconic Henry rifle, there is much more to this enigmatic individual than meets the eye. To truly understand his groundbreaking contributions to firearms technology, one must delve into his early life and influences, unearthing the formative experiences that shaped him into the legend he would become.

Born on March 22, 1821, in Claremont, New Hampshire, Benjamin Tyler Henry grew up in a time of great societal and technological change. Raised in an era when firearms were evolving rapidly from flintlock muskets to more advanced mechanisms like percussion cap rifles, young Benjamin developed an early passion for engineering and innovation. His natural curiosity led him to spend countless hours tinkering with machinery and exploring new ways to improve existing designs.

Henry's upbringing was influenced by his father's trade as a gunsmith. As a young apprentice in his father's shop, he learned invaluable skills that would lay the foundation for his future achievements. By observing firsthand how guns were manufactured and repaired, he gained practical knowledge of firearm mechanics that would prove essential later on.

However, it was Henry's encounter with Horace Smith during his teenage years that marked a turning point in his life. Smith—a skilled gunsmith himself—recognized Benjamin's remarkable talent and took him under his wing as an apprentice at the renowned Smith & Wesson factory in Norwich, Connecticut. Under Smith's mentorship, Henry honed his skills further while also gaining exposure to cutting-edge manufacturing techniques.

The experiences gained during this period deeply influenced Henry's approach to firearms design. Inspired by innovative concepts such as interchangeable parts pioneered by Eli Whitney Jr., he recognized their potential application within firearm manufacturing—a radical departure from traditional gunsmithing methods prevalent at the time.

Henry's exposure to emerging technologies and his inquisitive nature led him to experiment with various ideas. It was during this period that he first conceived the concept of a lever-action repeating rifle, a revolutionary idea that would forever change the landscape of firearms.

Beyond his technical skills and influences, Henry's personal life played a significant role in shaping his character. Known for his unwavering determination and tireless work ethic, he faced numerous setbacks throughout his career but always managed to persevere. This resilience can be traced back to the values instilled in him by his family—hard work, integrity, and a commitment to excellence.

Benjamin Tyler Henry's early life and influences provide crucial insights into the man behind the legend. His formative experiences as an apprentice gunsmith and exposure to innovative technologies set him on a path of groundbreaking firearm design. Combined with personal traits such as determination and resilience, these influences propelled Henry towards creating one of the most iconic firearms in American history—the Henry rifle.

Truly understanding this remarkable individual requires uncovering the layers beneath the legend—a journey that reveals an extraordinary man driven by passion and ingenuity.

***

In the annals of American history, few names resonate as strongly as Benjamin Tyler Henry. Revered as a pioneer in the field of firearms manufacturing, Henry's contributions forever changed the landscape of warfare and hunting. However, it was the untamed and rugged American West that served as a catalyst for his innovations.

Born in 1821, Benjamin Tyler Henry grew up during a time when the United States was rapidly expanding westward. As settlers ventured into new territories, they faced numerous challenges and dangers that demanded reliable weaponry. It was against this backdrop that Henry's inventive spirit thrived.

One cannot fully understand the genesis of Benjamin Tyler Henry's innovations without acknowledging the significance of his time spent in Kansas during its tumultuous Bleeding Kansas era. The region was embroiled in violent clashes between pro-slavery and anti-slavery factions, creating an urgent need for effective firearms capable of defending one's life and property.

It was during this turbulent period that Henry conceptualized his most famous creation – the lever-action repeating rifle. Driven by a desire to provide settlers with superior firepower, he devised a mechanism that allowed rapid firing without requiring manual reloading after each shot.

Henry's lever-action design proved to be a game-changer on the Western frontier. Its unparalleled speed and reliability made it a favorite among soldiers, lawmen, and outlaws alike. With its ability to hold up to sixteen rounds in its tubular magazine – an astonishing feat at the time – this firearm became known as "the rifle you could load on Sunday and shoot all week."

The vast expanses of untamed wilderness in the American West presented another crucial factor that spurred Henry's innovative drive: hunting. As pioneers ventured into uncharted territories teeming with wildlife, they required firearms capable of efficiently taking down large game while still being lightweight enough to carry over long distances.

Henry rose to the challenge by creating a lighter version of his lever-action rifle, specifically tailored for hunting purposes. This firearm quickly gained popularity among frontiersmen, who relied on it to provide food for their families and protect themselves from wild animals.

The American West acted as a proving ground for Benjamin Tyler Henry's revolutionary firearms. The harsh conditions and constant threats faced by settlers demanded weaponry that was not only effective but also durable and easy to maintain. Henry's designs excelled in all these aspects, earning him widespread acclaim and establishing his name as synonymous with excellence in firearms manufacturing.

As the American West evolved, so did Henry's innovations. His firearms played an integral role during the Civil War, aiding both Union soldiers and Confederate troops. Furthermore, they continued to shape the course of history as pioneers pushed further into the frontier, taming the wilderness one shot at a time.

It is undeniable that Benjamin Tyler Henry's innovations were profoundly influenced by his experiences in the American West. The need for reliable firearms on the frontier propelled him to create groundbreaking designs that revolutionized warfare and hunting alike. Today, his legacy lives on as his name remains firmly etched in history books as one of America's greatest inventors and pioneers.

During the 19th century, the United States experienced a rapid westward expansion that forever changed the landscape and culture of the nation. As pioneers ventured further into uncharted territories, they faced numerous challenges and dangers, which led to an increasing demand for firearms that were both reliable and efficient. At the forefront of this revolution in weaponry was Benjamin Tyler Henry, whose innovative designs would shape the course of American history.

The westward expansion brought settlers face to face with harsh environments, untamed wilderness, and hostile Native American tribes. These circumstances necessitated the use of firearms as tools for survival and defense. However, traditional firearms at the time were often unreliable, slow to reload, and lacked accuracy over long distances. This prompted a need for revolutionary firearms capable of meeting these challenges head-on.

Benjamin Tyler Henry recognized this pressing demand and set out to create a firearm that would revolutionize the industry. In 1860, he introduced his most famous creation: The Henry Repeating Rifle. Unlike its predecessors, this lever-action rifle had a tube magazine located under its barrel that could hold up to fifteen rounds of ammunition – an impressive capacity for its time.

This design allowed users to rapidly fire multiple shots without needing to reload after each shot.

The Henry Repeating Rifle became synonymous with Western expansion due to its exceptional performance in real-world situations. Its quick rate of fire gave settlers a distinct advantage when facing threats such as attacks from Native American tribes or encounters with dangerous wildlife like bears or mountain lions. The rifle's reliability also ensured that it could withstand harsh weather conditions prevalent in many western territories.

One notable example highlighting the significance of Henry's firearm is its impact on conflicts during this era. During the American Civil War (1861-1865), both Union soldiers and Confederate troops utilized various models of Henry rifles due to their unmatched firepower and reliability on the battlefield. Soldiers armed with these revolutionary firearms often had a significant advantage over their adversaries, as they could fire multiple rounds accurately before their enemies had a chance to reload.

Furthermore, the Henry Repeating Rifle was instrumental in transforming the concept of warfare during this period. Its rapid-fire capabilities challenged traditional military tactics, forcing commanders to adapt and develop new strategies. The rifle's effectiveness also led to increased demand from both military forces and civilian settlers, further fueling Western expansion.

Western expansion in the 19th century created an urgent need for revolutionary firearms capable of overcoming the challenges faced by pioneers and settlers. Benjamin Tyler Henry's innovative designs, particularly his famous creation – the Henry Repeating Rifle – forever altered the course of American history. Its exceptional performance in real-world situations and its impact on conflicts during this era solidified its position as an iconic symbol of Western expansion and a testament to Benjamin Tyler Henry's remarkable contributions to firearm technology.

Benjamin Tyler Henry, a name synonymous with innovation and revolution in the world of firearms, is best known for his groundbreaking invention - the lever-action rifle. This remarkable firearm not only changed the course of history but also solidified Henry's place as one of the greatest inventors in American firearms technology.

During the mid-19th century, single-shot muzzle-loading rifles were prevalent, but they were slow and cumbersome to reload. This drawback led Benjamin Tyler Henry to envision a weapon that would provide rapid-fire capability without compromising accuracy or reliability. His visionary concept gave birth to what would become an iconic firearm design.

The lever-action rifle designed by Henry was a marvel of engineering and ingenuity. It featured a tubular magazine beneath the barrel, allowing it to hold multiple rounds of ammunition. By operating a lever located near the trigger guard, shooters could easily feed cartridges from the magazine into the chamber, readying their next shot in a fraction of the time it took with traditional rifles.

One of the key innovations that made Henry's lever-action rifle so game-changing was its smooth and efficient action mechanism. The lever itself operated both as a means to cycle rounds into position and as a trigger guard, eliminating unnecessary movements during reloading. This mechanism significantly increased firing speed compared to other contemporary firearms.

The introduction of this innovative rifle had far-reaching consequences for military tactics and personal defense alike. Its rapid-fire capability allowed soldiers on both sides during the American Civil War to unleash devastating volleys upon their enemies swiftly. The firepower advantage provided by these rifles undoubtedly influenced battlefield strategies and helped shape modern warfare.

Beyond military application, Benjamin Tyler Henry's invention found immense popularity among civilians seeking reliable firearms for hunting or self-defense purposes. The ease of use and quick reloading made it ideal for those facing dangerous situations where speed was paramount.

Henry's design didn't just revolutionize firearms; it also paved the way for subsequent advancements in firearm technology. The lever-action rifle became the foundation for iconic firearms such as the Winchester Model 1866 and 1873, both renowned for their reliability, accuracy, and versatility.

The legacy of Benjamin Tyler Henry's lever-action rifles can still be felt today. While modern semi-automatic firearms have largely replaced lever-action rifles in military use, they remain highly sought-after by collectors and enthusiasts worldwide. The enduring popularity of these firearms stands as a testament to Henry's genius and his contribution to the evolution of firearms technology.

Benjamin Tyler Henry's lever-action rifle forever changed the landscape of firearms with its rapid-fire capability, efficient action mechanism, and ease of use. Its impact on military tactics and personal defense cannot be overstated. This revolutionary invention is a testament to Henry's brilliance as an inventor and solidifies his place in history as one of the greatest minds behind American firearms technology.

The American West during the mid-19th century was a vast and untamed frontier, where pioneers and settlers faced numerous challenges. Among these challenges was the need for reliable and efficient firearms to protect themselves, hunt for sustenance, and defend their homesteads against Native American raids. Enter Benjamin Tyler Henry, an ingenious gunsmith whose lever-action rifles would forever change the course of history in the American West.

Henry's lever-action rifles were a revolutionary advancement in firearms technology that offered unparalleled firepower and reliability. Prior to Henry's innovations, most firearms utilized single-shot mechanisms or slow-loading muzzleloaders. These weapons were laborious to operate, often requiring individuals to stand still while reloading, leaving them vulnerable in dangerous situations.

In contrast, Henry's lever-action rifles featured a tubular magazine located beneath the barrel that could hold multiple rounds of ammunition. The lever action allowed for quick and seamless cycling of cartridges, enabling users to fire multiple shots rapidly without needing to reload after each shot. This game-changing feature gave pioneers and settlers an immense advantage in terms of firepower when facing threats in the Wild West.

One significant impact of Henry's lever-action rifles was their effectiveness in hunting game animals for sustenance. The abundance of wildlife in the American West provided a vital food source for settlers, but traditional firearms were often inadequate for taking down large or fast-moving animals such as bison or deer. With its rapid-fire capabilities and increased ammunition capacity, Henry's rifle became an essential tool for successful hunting expeditions.

Moreover, Henry rifles played a pivotal role in securing homesteads against raids by Native American tribes who fiercely defended their ancestral lands from encroachment by settlers. The superior firepower offered by these weapons gave settlers a significant advantage during conflicts with native populations who relied on older muzzleloading firearms or bows and arrows.

The impact of Henry's lever-action rifles extended beyond practical applications; it also played a vital role in shaping the perception and mythology of the American West. These rifles became iconic symbols of frontier life, featured prominently in popular literature, dime novels, and later in Hollywood westerns. The image of a cowboy riding into the sunset with a Henry rifle slung across his back became an enduring symbol of American individualism and the rugged spirit of the West.

Benjamin Tyler Henry's lever-action rifles had a profound impact on the American West. Their rapid-fire capabilities and increased ammunition capacity revolutionized firearms technology, providing pioneers and settlers with unparalleled firepower for hunting, self-defense, and securing their homesteads. Moreover, these rifles played a significant role in shaping the mythology surrounding the American West. Henry's firearms remain an enduring testament to his ingenuity and continue to captivate enthusiasts to this day.

***

In the mid-19th century, as the United States embarked on a period of rapid westward expansion, firearms played a pivotal role in shaping the course of this historic journey. Among the numerous firearms that emerged during this time, one name stands out: Benjamin Tyler Henry and his revolutionary rifles. The introduction of Henry's firearms not only transformed warfare but also influenced settlement patterns and economic development, and ultimately contributed to the triumph of American exceptionalism.

The timing of Henry's firearms could not have been more opportune. As settlers moved westward, they encountered vast landscapes populated by Native American tribes fiercely protective of their territories. These encounters often led to conflicts that required superior firepower for survival. The traditional single-shot muzzleloaders used by both sides were no match for the devastating firepower offered by Henry's repeating rifles.

Henry's rifle was unlike anything seen before. Its innovative design incorporated a magazine tube that held up to 16 metallic cartridges, allowing for rapid fire without the need for reloading after each shot. This breakthrough in firearm technology gave settlers and soldiers a distinct advantage over their adversaries while embarking on their perilous journeys into unknown territories.

The impact of Henry's firearms on western expansion cannot be overstated. The ability to fire multiple rounds without reloading made it possible to hold off attacks from larger numbers or mounted enemies, such as Native American warriors on horseback, effectively. This increased firepower drastically altered battle dynamics and provided settlers with newfound confidence and security as they ventured into untamed lands.

Furthermore, Henry's rifles played an instrumental role in fueling economic development throughout the western expansion. Settlers armed with these advanced weapons could defend their livestock from predators or thieves more efficiently than ever before. Additionally, hunters armed with Henry rifles were able to procure larger quantities of game for food and fur trade purposes, thus stimulating local economies as well as supplying valuable resources needed for survival in the wilderness.

The widespread adoption of Henry's firearms also contributed to the concept of American exceptionalism. As pioneers armed with these powerful rifles successfully defended their claims, they not only solidified their position on the frontier but also reinforced the notion that America was destined to conquer and tame this vast continent. The stories of settlers armed with Henry rifles fending off attacks from Native Americans or outlaws became legendary tales that fueled national pride and further motivated westward migration.

Benjamin Tyler Henry's revolutionary firearms played a pivotal role in shaping the course of Western expansion during the mid-19th century. By providing settlers and soldiers with unprecedented firepower, these repeating rifles transformed warfare influenced settlement patterns, stimulated economic development, and contributed to the triumph of American exceptionalism. The legacy of Henry's firearms endures as a testament to their profound impact on this critical period in American history.

Benjamin Tyler Henry, a name that resonates throughout history as the man behind the legendary Henry rifle, left an indelible mark on the world of firearms. His revolutionary design not only revolutionized the way wars were fought but also played a pivotal role in shaping the American frontier. Today, his legacy lives on as we celebrate his immense contributions to history.

Henry's invention of the lever-action repeating rifle in 1860 forever changed the face of warfare. Unlike its predecessors, which required manual reloading after each shot, Henry's rifle could fire multiple rounds without pausing to reload. This groundbreaking advancement gave soldiers an unprecedented advantage on the battlefield, allowing them to unleash a hailstorm of bullets upon their enemies. The effectiveness and reliability of this firearm quickly gained recognition and earned it high praise among military ranks.

During the American Civil War (1861-1865), Union soldiers armed with Henry rifles became formidable opponents of Confederate forces. Their ability to rapidly fire shots gave them a significant advantage in battles such as Gettysburg and

Chickamauga. General Ulysses S. Grant himself recognized this remarkable weapon's impact on war strategy when he famously stated that "a regiment or brigade armed with it can face and successfully resist ten times its number armed with muzzle-loaders."

Beyond military applications, Benjamin Tyler Henry's contribution extended into civilian life, especially during westward expansion in America. Settlers venturing into untamed territories relied heavily on firearms for hunting game or defending against potential threats from Native American tribes or bandits. The Henry rifle quickly became a symbol of power and protection for these pioneers, securing its place as an iconic piece of Americana.

Despite his undeniable impact on history, it is unfortunate that Benjamin Tyler Henry did not receive proper recognition during his lifetime. While he did patent his invention and established the New Haven Arms Company (later renamed the Winchester Repeating Arms Company), financial difficulties plagued him, and he ultimately sold his patent rights. However, Henry's contributions were not forgotten.

Today, historians and firearm enthusiasts alike recognize Benjamin Tyler Henry as a true pioneer in the field of firearms design. His innovative lever-action mechanism paved the way for subsequent advancements in rifle technology, influencing generations of gunsmiths and manufacturers. The Henry rifle's legacy endures through its continued production by the modern-day Henry Repeating Arms Company, which stays true to its founder's vision of crafting exceptional firearms.

In recognition of his immense contributions to history, Benjamin Tyler Henry was posthumously inducted into the National Inventors Hall of Fame in 2019. This prestigious honor acknowledges his innovative spirit and lasting impact on American culture. Additionally, museums across the country proudly display original Henry rifles as prized artifacts, ensuring that future generations can appreciate and learn from this remarkable piece of history.

Benjamin Tyler Henry's legacy stretches far beyond his lifetime. His invention revolutionized warfare and played a significant role in shaping America's frontier era. As we celebrate his contributions to history, let us remember him as a visionary whose revolutionary firearms forever changed the world we live in today.

***

The name Benjamin Tyler Henry is synonymous with one of the most iconic firearms in American history, the Henry rifle. Revered as a revolutionary weapon during the Civil War era, it forever changed the course of firearms development. Yet, while the rifle has received much attention and recognition, the man behind its creation has often remained a forgotten figure in history.

In recent years, however, renewed interest and research have allowed us to uncover forgotten chapters of Benjamin Tyler Henry's remarkable story.

Born on March 22, 1821, in Claremont, New Hampshire, Benjamin Tyler Henry displayed an early aptitude for mechanics and engineering. As a young man working at Robbins & Lawrence Armory in Windsor, Vermont, he honed his skills and gained invaluable experience in firearm manufacturing. This period would prove pivotal for his future endeavors.

In 1850, Henry moved to New Haven and joined forces with Oliver Winchester to establish the New Haven Arms Company (later renamed Winchester Repeating Arms Company). It was here that he began developing what would become his most significant contribution to firearms technology – the lever-action repeating rifle.

While many associate Henry solely with his eponymous rifle model introduced during the Civil War era, it is important to note that this was not his only innovation. Recent research has shed light on several lesser-known inventions by Henry that had a profound impact on firearm design.

Among these innovations was an improved breech-loading mechanism that allowed for quicker reloading times compared to traditional muzzle-loading rifles. Prior to this breakthrough invention by Henry in 1854-1855, breech-loading firearms were prone to gas leakage and required complex mechanisms. His design effectively addressed these issues by using a simple hinged-block mechanism that was sealed tightly upon firing.

Furthermore, Benjamin Tyler Henry's contributions extended beyond firearms technology itself. He was an astute businessman who recognized the importance of marketing and branding. Henry actively promoted his firearms through advertising campaigns, ensuring that the name "Henry" became synonymous with quality and reliability.

Sadly, despite his remarkable achievements and contributions to the firearms industry, Benjamin Tyler Henry's life ended tragically. He passed away on December 8, 1898, in New Haven, Connecticut. However, the legacy he left behind continues to shape the world of firearms to this day.

Rediscovering Benjamin Tyler Henry has allowed us to unravel forgotten chapters of his story. Beyond being known solely as the inventor of the iconic Henry rifle, we now recognize him as a brilliant engineer who made several crucial contributions to firearm design. Through his innovative breech-loading mechanism and astute marketing strategies, Henry left an indelible mark on both technological advancements and business practices in the firearms industry.

As we continue to explore his life and work further, we gain a deeper appreciation for this visionary figure who revolutionized firearms manufacturing forever.

***

Throughout history, there have been countless individuals whose contributions have shaped the world we live in today. Benjamin Tyler Henry, the man behind the legend of revolutionary firearms, undoubtedly belongs among these pioneers. His groundbreaking designs and relentless pursuit of excellence propelled him to the forefront of firearm innovation during his time and left a lasting legacy that continues to resonate with enthusiasts and professionals alike.

Henry's unwavering commitment to developing firearms that were not only reliable but also highly efficient revolutionized the industry. His most renowned invention, the Henry rifle, was a game-changer. It incorporated an innovative design featuring a tubular magazine under the barrel, allowing for rapid and continuous fire without needing to reload after every shot. This breakthrough technology elevated Henry's rifles far above their competitors and set new standards for efficiency and effectiveness in firearms.

Beyond his technical achievements, Henry's impact extended beyond the realm of firearms themselves. He played an integral role in shaping American history during critical moments such as the Civil War. The Union Army's adoption of the Henry rifle provided them with a significant advantage over Confederate forces due to its rapid-fire capabilities. This advantage not only saved countless lives but also played a pivotal role in securing victory for the Union.

Furthermore, Benjamin Tyler Henry demonstrated his entrepreneurial spirit by founding The New Haven Arms Company (later renamed The Winchester Repeating Arms Company). Under his leadership as its superintendent, he continued to push boundaries in firearm design and manufacture long after leaving his mark with the Henry rifle. His commitment to innovation was evident through subsequent models like the famous Winchester Model 1866 lever-action rifle – often referred to as "The Yellow Boy" due to its brass frame – which further solidified his reputation as an innovator.

Today, enthusiasts continue to celebrate Benjamin Tyler Henry's contributions by collecting and cherishing these iconic firearms that bear his name. The legacy he left behind through his pioneering designs has not only stood the test of time but has also inspired generations of inventors and firearm enthusiasts alike. His impact on the industry cannot be overstated, and his work continues to shape the future of firearms.

Benjamin Tyler Henry's story is one that deserves recognition and appreciation. Through his revolutionary firearms and unwavering dedication to excellence, he forever changed the landscape of firearm design and left an indelible mark on American history. As we honor his memory, let us remember not only the man behind the legend but also the countless lives he influenced through his innovations.

Benjamin Tyler Henry's legacy will continue to inspire future generations of inventors, ensuring that his pioneering spirit lives on in the ever-evolving world of firearms.

Wilhelm Mauser
1834-1882

Wilhelm Mauser, the co-founder of the renowned firearms manufacturer Mauser, was a visionary engineer whose contributions revolutionized the world of firearms. Born on May 2, 1834, in Oberndorf am Neckar, Germany, Wilhelm's early life laid the foundation for his exceptional talent and passion for engineering.

Growing up in a family deeply rooted in gunsmithing traditions, it was only natural that Wilhelm would be drawn to this craft. His father, Franz Andreas Mauser, owned a small workshop where he crafted rifles and pistols. From an early age, Wilhelm would spend countless hours observing his father's work and developing a keen understanding of firearm mechanics.

However, tragedy struck when Wilhelm was just nine years old. His father passed away suddenly due to illness. This unfortunate event forced young Wilhelm to take up responsibilities beyond his years. With his mother struggling to make ends meet after her husband's death, young Wilhelm had to combine his studies with part-time jobs at local workshops.

Despite these challenges, Wilhelm's determination only grew stronger. He continued honing his skills by working alongside master craftsmen and attending technical schools during evenings and weekends. These experiences exposed him to different approaches in engineering and fueled his curiosity for innovation.

In 1859, at the age of 25, Wilhelm joined forces with his older brother Paul Mauser to establish their own company - "Gebrüder (Brothers) Mauser." This marked the beginning of their journey towards greatness as they embarked on creating superior firearms that would redefine industry standards.

Wilhelm's genius soon became evident through several groundbreaking inventions that shaped modern firearm design. One such innovation was the introduction of metallic cartridges instead of traditional muzzle-loading systems. This invention revolutionized not only their company but also military arsenals around the world.

His relentless pursuit of perfection led him to develop the famous Mauser bolt-action rifle, which set new benchmarks for accuracy, reliability, and ease of use. This masterpiece earned immense recognition and became the foundation for future generations of firearms.

Wilhelm's expertise in engineering extended beyond firearms. He also played a pivotal role in developing machinery for manufacturing ammunition, ensuring the consistent quality and availability of their products.

Beyond his technical contributions, Wilhelm's leadership and business acumen were instrumental in transforming Mauser into a global brand. Under his guidance, the company expanded its reach to international markets, establishing partnerships with governments and private entities worldwide.

Wilhelm Mauser's life was tragically cut short when he passed away on January 13, 1882. However, his legacy lives on through his remarkable contributions to firearm innovation. His relentless pursuit of perfection and dedication to craftsmanship continue to inspire engineers and designers across industries.

Today, Wilhelm Mauser is recognized as one of the greatest pioneers in firearm engineering. His unwavering commitment to excellence has left an indelible mark on the industry, making him an icon whose genius will forever be celebrated.

***

Firearms have played a significant role in human history, revolutionizing warfare, hunting, and personal protection. From ancient times to the modern era, the evolution of firearms has been marked by numerous innovations and advancements. Among the many brilliant minds that have contributed to this progression, Wilhelm Mauser stands out as a true genius behind Mauser firearms. The history of firearms can be traced back to ancient China, where the invention of gunpowder in the 9th century laid the foundation for early explosive devices.

The first true firearms appeared in Europe during the 14th century as crude muzzle-loading cannons and hand cannons. These early weapons were cumbersome and unreliable, but they marked an important turning point in military technology. The 16th century witnessed significant advancements in firearm design with the introduction of matchlock mechanisms. This innovation allowed for more accurate firing by utilizing a slow burning matchcord to ignite gunpowder.

Matchlock muskets quickly became widespread across Europe and shaped warfare for centuries to come. Innovation continued into the 17th century when flintlock mechanisms replaced matchlocks. Flintlocks utilized a piece of flint striking against steel to produce sparks that ignited gunpowder. This improvement made firearms more reliable and easier to use while maintaining accuracy. During the 19th century, breech-loading firearms emerged as a game-changer in firearm design.

Instead of loading from the muzzle end like their predecessors, these guns allowed ammunition to be loaded from the rear or side of the barrel, significantly reducing reloading time. The development of rifling also greatly improved accuracy by introducing spiral grooves inside barrels that imparted spin on bullets. It was within this backdrop of rapid technological advancements that Wilhelm Mauser made his mark on firearm history.

Born in Germany in 1834, Mauser began his career working alongside his older brother, Paul Mauser. Together, they founded the Mauser company, which would become synonymous with excellence in firearm design. Wilhelm Mauser's innovations focused on improving reliability, safety, and accuracy. One of his most significant contributions was the development of the bolt-action rifle system. The Mauser Model 1871 rifle introduced a robust bolt mechanism that locked securely into the breech, ensuring consistent accuracy and preventing dangerous malfunctions.

Building upon this success, Wilhelm Mauser continued to refine and improve his designs. The introduction of the famous Mauser 98 rifle in 1898 solidified his reputation as one of the greatest firearm designers of all time. This rifle incorporated features such as controlled feeding, a strong and reliable extractor, and smooth bolt operation that set new standards for military rifles.

Wilhelm Mauser's legacy extends far beyond his lifetime. His designs formed the basis for many later firearms used by militaries around the world. Even today, over a century later, Mauser firearms are sought after by collectors and enthusiasts who appreciate their timeless elegance and exceptional performance. Understanding the history of firearms is crucial to appreciating Wilhelm Mauser's remarkable contributions to this field.

***

In the realm of firearms, few names stand as tall as Wilhelm Mauser. The German gunsmith and engineer revolutionized military weaponry with his groundbreaking innovations, particularly in the development of bolt-action rifles. Through his relentless pursuit of perfection and meticulous attention to detail, Mauser forever changed the landscape of modern firearms, leaving an indelible mark on military history.

During the late 19th century, bolt-action rifles were gaining popularity due to their superior accuracy and rapid-fire capabilities. However, it was Wilhelm Mauser who refined and perfected this design, elevating bolt-action rifles to new heights of performance. His tireless dedication to improving firearm technology led him to create a series of iconic rifles that would soon become legendary.

One such rifle was the Gewehr 71, introduced in 1871 as Germany's first modern infantry rifle. This revolutionary weapon featured a magazine-fed bolt action system that greatly enhanced its reloading speed compared to its predecessors. The Gewehr 71 marked a significant departure from traditional muzzle-loading rifles and set the stage for future advancements in military weaponry.

Building upon the success of the Gewehr 71, Mauser went on to develop an even more remarkable firearm—the Gewehr 98. Introduced in 1898, this bolt-action rifle became widely regarded as one of the finest military rifles ever made. Its innovative design incorporated several key features that improved accuracy, reliability, and ease of use.

One groundbreaking feature was Mauser's patented "controlled-feed" system which ensured reliable cartridge feeding by capturing and controlling each round during cycling—a crucial improvement over previous designs prone to jamming under intense combat conditions. Additionally, the Gewehr 98 boasted a five-round internal magazine and a powerful cartridge that extended its effective range beyond what had been previously achieved.

The influence of Wilhelm Mauser's designs extended far beyond Germany's borders. The Gewehr 98, in particular, became the blueprint for many subsequent bolt-action rifles used by various nations around the world. Its success and widespread adoption solidified Mauser's position as a visionary firearms engineer.

Moreover, Mauser's impact on military weaponry was not limited to his groundbreaking designs alone. His emphasis on precision manufacturing and quality control set new industry standards that endure to this day. The meticulous craftsmanship and attention to detail showcased in each Mauser rifle elevated the art of gun making, ensuring that these firearms were not just functional tools but works of art.

The rise of bolt-action rifles was undoubtedly a watershed moment in military history, forever altering the way wars were fought. Wilhelm Mauser's relentless pursuit of perfection and his unwavering commitment to innovation changed military weaponry forever. His iconic rifles continue to be revered by collectors and enthusiasts alike, standing as a testament to his genius and indelible mark on the firearms industry.

***

Wilhelm Mauser, a name synonymous with excellence and innovation in the world of firearms, left an indelible mark on the industry through his groundbreaking designs and revolutionary engineering. As we celebrate the genius behind Mauser firearms, it is crucial to delve into the key contributions that Wilhelm Mauser made to firearms design, forever changing the landscape of weapon manufacturing.

One of Wilhelm Mauser's most significant contributions was his development of the bolt-action rifle mechanism. Prior to his invention, most rifles were equipped with single-shot or muzzle-loading mechanisms that required manual reloading after each shot. However, Mauser's ingenious design incorporated a bolt action system that allowed for rapid and efficient loading and firing. This breakthrough not only increased the overall rate of fire but also improved accuracy and reliability, setting new standards for firearm performance.

Another milestone in Mauser's illustrious career was his creation of a detachable box magazine. By introducing this innovation, he eliminated the need for individual loading cartridges into a firearm manually. The detachable box magazine allowed for quick reloads by simply swapping out an empty magazine with a fully loaded one. This advancement revolutionized military tactics as soldiers could now engage in prolonged firefights without interrupting their shooting rhythm.

Furthermore, Wilhelm Mauser was instrumental in developing smokeless powder cartridges. Before his time, black powder was used as a propellant in ammunition; however, it produced copious amounts of smoke upon ignition, limiting visibility on the battlefield. Recognizing this drawback, Mauser sought to create an alternative propellant that would eliminate smoke production while maintaining similar or improved ballistic properties. His efforts resulted in smokeless powder cartridges that enhanced accuracy while providing soldiers with better concealment during combat operations.

Additionally, Mauser played an integral role in perfecting barrel rifling techniques. Through meticulous experimentation and refinement processes, he developed innovative methods of rifling the barrels, significantly enhancing the projectile's stability and accuracy. This breakthrough not only advanced the performance of Mauser firearms but also influenced future designs across the industry.

Moreover, Wilhelm Mauser's unwavering commitment to quality and precision manufacturing cannot be overlooked. His meticulous attention to detail ensured that every Mauser firearm produced was a testament to exceptional craftsmanship. This dedication to excellence became a hallmark of Mauser firearms, earning them a reputation for reliability and durability that remains unparalleled even today.

Wilhelm Mauser's genius and innovative spirit forever changed the landscape of firearms design. His contributions, including the bolt-action rifle mechanism, detachable box magazine, smokeless powder cartridges, and advancements in barrel rifling techniques, revolutionized weapon manufacturing and military tactics. Furthermore, his commitment to quality set new standards for craftsmanship in the industry. As we celebrate Wilhelm Mauser's enduring legacy, we must recognize his key contributions as cornerstones of modern firearms design that continue to shape the industry today.

***

The birth of the Mauser system marked a significant milestone in the history of firearms, revolutionizing bolt-action rifle technology and setting a new standard for precision, reliability, and versatility. At the forefront of this revolution was Wilhelm Mauser, a brilliant German engineer who dedicated his life to perfecting firearms design.

In the late 19th century, bolt-action rifles were already in use, but they often suffered from issues such as jamming and unreliable extraction. Wilhelm Mauser recognized these shortcomings and sought to develop a rifle that would address these problems while enhancing overall performance. His quest for perfection led to the creation of what would become known as the Mauser system.

Central to the Mauser system was its innovative bolt design. Unlike previous designs that relied on complex mechanisms prone to malfunctioning, Mauser's bolt featured a simple yet robust construction. The three-lug rotating bolt head provided increased strength and stability during firing while ensuring smooth operation. This design not only improved accuracy but also greatly enhanced reliability by minimizing jamming and extraction issues.

Another crucial feature of the Mauser system was its integral box magazine. Prior to this innovation, most rifles used external magazines or relied on single shot loading mechanisms. By incorporating an internal magazine with a capacity for multiple rounds, Wilhelm Mauser made reloading faster and more efficient, enabling soldiers or hunters to fire several shots without having to reload after each shot manually.

Furthermore, the introduction of controlled-feed feeding mechanism in the Mauser system eliminated another prevalent issue with earlier designs – cartridge misalignment during chambering. This feature ensured reliable feeding even under adverse conditions such as rapid firing or when handling cartridges with varying dimensions.

The genius behind Wilhelm Mauser's innovations extended beyond functional improvements; he also paid great attention to ergonomics. The design featured an ergonomic stock shape that fit comfortably into a shooter's shoulder pocket, reducing recoil felt by the shooter while allowing for quick target reacquisition after each shot. The Mauser system also incorporated a straight-line stock design, reducing muzzle rise and enhancing overall stability during firing.

The versatility of the Mauser system further contributed to its success. By offering modular components, it allowed for easy customization and adaptation to various calibers, barrel lengths, and stock configurations. This versatility made the Mauser system suitable for military use as well as civilian applications such as hunting or target shooting.

Wilhelm Mauser's genius behind the birth of the Mauser system revolutionized bolt-action rifle technology forever. Through his relentless pursuit of perfection, he created a firearm that combined precision, reliability, and versatility like never before. The innovative bolt design, integral box magazine, controlled feed feeding mechanism, ergonomic features, and modular components set new standards in the industry. Today, more than a century later, the legacy of Wilhelm Mauser lives on in the countless firearms that bear his name and continue to be celebrated for their exceptional quality and performance.

***

Wilhelm Mauser, along with his brother Paul, revolutionized the firearms industry with their ingenuity and relentless pursuit of perfection. Under Wilhelm's leadership, Mauser firearms underwent a remarkable evolution, transforming from mere prototypes into iconic masterpieces that set new standards for reliability, accuracy, and functionality.

At the outset of their journey, Wilhelm and Paul Mauser recognized the need for innovation in firearm design. They sought to develop a weapon that would surpass all existing models in terms of performance and durability. The brothers began by experimenting with various designs and mechanisms until they arrived at what would become known as the Mauser bolt-action system.

The early prototypes showcased Wilhelm's dedication to precision engineering. He meticulously refined each component, ensuring optimal functionality and ease of use. The bolt-action mechanism allowed for quicker reloading times while maintaining exceptional accuracy—a feature that would become synonymous with Mauser firearms.

As word spread about the exceptional performance of their prototypes, demand grew rapidly. Recognizing this opportunity for growth, Wilhelm embraced an iterative approach to design. He welcomed feedback from users and incorporated their suggestions into subsequent iterations—an early example of user-centered design.

One significant milestone in the evolution of Mauser firearms was the introduction of smokeless powder cartridges. This advancement provided higher velocity without compromising safety or reliability—a testament to Wilhelm's commitment to technological progress. With these modern cartridges, Mauser rifles became renowned for their long-range accuracy and stopping power on the battlefield.

Wilhelm's passion for continuous improvement led him to explore novel manufacturing techniques as well. He introduced mass production methods that allowed for increased efficiency without compromising quality control—another pioneering move within the industry.

The culmination of this tireless pursuit came with the introduction of one of the most iconic firearms ever created—the Mauser Model 98 rifle. Launched in 1898, it quickly gained global recognition as the epitome of bolt-action rifles. The Model 98 boasted exceptional accuracy, reliability, and versatility, making it a favorite among military forces and hunters alike.

Wilhelm's leadership extended beyond firearm design. He recognized the importance of marketing and establishing a strong brand identity. By actively promoting Mauser firearms at international exhibitions and competitions, he solidified their reputation as the gold standard in firearms.

The legacy of Wilhelm Mauser's innovations lives on to this day. Countless manufacturers have drawn inspiration from his designs, ensuring that his contributions continue to shape modern firearms. From hunting rifles to military arms, the influence of Wilhelm's relentless pursuit of perfection can be seen in every aspect of firearm design.

Wilhelm Mauser's leadership was instrumental in transforming Mauser firearms from prototypes into renowned masterpieces. His dedication to precision engineering, continuous improvement, and technological advancement set new benchmarks for the industry. Today, we celebrate Wilhelm's genius and his enduring impact on the world of firearms.

***

The impact of Mauser bolt-action rifles on military tactics was profound, shaping warfare strategies and revolutionizing the way armies fought on the battlefield. Wilhelm Mauser's innovations in firearm design not only enhanced the capabilities of individual soldiers but also influenced broader military doctrines and tactics.

One of the key aspects that set Mauser rifles apart was their exceptional accuracy and range. The introduction of smokeless powder cartridges in Mauser rifles allowed for greater muzzle velocity, extending effective firing ranges. This increased range forced armies to reconsider their traditional tactics, which had relied heavily on close-quarters combat.

Mauser bolt-action rifles facilitated rapid fire rates, enabling soldiers to engage multiple targets quickly. The ability to reload swiftly with a bolt action mechanism gave infantrymen an advantage over adversaries using slower-to-reload weapons like single-shot or lever-action rifles. This enhanced firepower significantly impacted military tactics as it allowed troops to suppress enemy positions effectively and maintain a sustained rate of fire during engagements.

Moreover, the Mauser's five-round internal magazine introduced an element of superiority over other firearms with smaller capacities. This feature enabled soldiers to engage multiple targets without having to reload as frequently, giving them a considerable advantage in battles where ammunition supply might be limited or difficult to access.

The accuracy and reliability of Mauser rifles led to advancements in marksmanship training within militaries worldwide. Soldiers armed with these weapons could engage targets at longer distances than ever before, necessitating more extensive training programs for infantrymen. The development of snipers became crucial on the battlefield due to the exceptional precision offered by Mauser firearms. These sharpshooters played a critical role in disrupting enemy operations and creating fear among opposing forces.

The success and reputation of Mauser bolt-action rifles influenced tactical doctrines across different nations' armies. Recognizing the advantages offered by these firearms, military strategists began implementing new formations and maneuvers that optimized their capabilities fully. For example, "fire and movement" became a prominent tactic, where soldiers would provide covering fire while others advanced toward the enemy. This tactic relied on the accuracy and quick reload times of Mauser rifles to effectively suppress enemy positions while maintaining forward momentum.

The impact of Mauser rifles extended beyond individual soldiers and affected the overall organization of armies. The increased range and accuracy led to a shift from linear formations to more dispersed ones, reducing vulnerability to concentrated enemy fire. This change in formation allowed for greater flexibility in maneuvering on the battlefield, as troops could exploit cover more effectively.

The innovations introduced by Wilhelm Mauser in bolt-action rifle design had a significant impact on military tactics and warfare strategies. The accuracy, range, firepower, and reliability of Mauser rifles revolutionized marksmanship training, influenced tactical doctrines, and shaped how armies fought on the battlefield. These firearms played a crucial role in changing traditional notions of warfare by extending effective firing ranges, enhancing soldier capabilities, and necessitating new formations and maneuvers.

***

While Mauser firearms are renowned for their exceptional performance on the battlefield, their legacy extends far beyond military applications. The genius behind Mauser firearms, Wilhelm Mauser, also recognized the potential of his creations in civilian and sporting contexts. This subtopic aims to celebrate the innovations of Wilhelm Mauser by exploring how his firearms have been utilized outside of warfare.

One area where Mauser firearms have found significant civilian use is in hunting. Wilhelm Mauser's meticulous attention to detail and focus on accuracy made his rifles highly sought after by hunters around the world. The bolt-action design of many

Mauser rifles allowed for quick follow-up shots and reliable performance in rugged environments. These qualities were particularly appreciated by big game hunters who required a dependable weapon capable of downing large animals with precision.

The versatility and reliability of these firearms cemented their place as trusted companions for hunters seeking to challenge themselves in pursuit of wild game.

In addition to hunting, sporting applications for Mauser firearms have also flourished over time. Competitive shooting sports such as long-range target shooting, and practical shooting competitions have benefited greatly from the accuracy and precision offered by these weapons. With their smooth actions, crisp triggers, and inherent accuracy, various models of Mauser rifles became prized possessions among sports shooters looking to push themselves to new limits.

Furthermore, collectors worldwide recognize the historical significance and craftsmanship behind Mauser firearms. The elegant designs coupled with unrivaled engineering excellence make them highly desirable pieces among firearm enthusiasts around the globe. Many individuals collect these guns not only as a testament to Wilhelm Mauser's ingenuity but also as tangible reminders of pivotal moments in history where these weapons played a vital role.

Beyond hunting, sports shooting, and collecting, some innovative individuals have even repurposed or modified certain models of Mauser firearms for other unique applications. For instance, some gunsmiths have transformed Mauser rifles into high-precision benchrest rifles, allowing for extreme accuracy in competitive shooting. Others have adapted these firearms for long-range varmint hunting, where precision and power are paramount. These examples demonstrate the adaptability and enduring legacy of Mauser firearms.

While the genius behind Mauser firearms is often associated with their battlefield prowess, Wilhelm Mauser's innovations have transcended military applications. From hunting to competitive shooting sports to collecting, these firearms have found a place in various civilian contexts. The meticulous craftsmanship, accuracy, and versatility of Mauser rifles have made them cherished companions for those seeking performance beyond the battlefield.

By celebrating Wilhelm Mauser's contributions in civilian and sporting realms, we honor his lasting impact on firearm design and innovation.

***

Wilhelm Mauser, the co-founder of Mauser Firearms, was an exceptional engineer and inventor whose contributions to modern firearm design continue to shape the industry today. His relentless pursuit of perfection, innovative designs, and commitment to quality established a legacy that has endured for over a century.

One of Wilhelm Mauser's most significant contributions was the development of the bolt-action rifle. In collaboration with his brother Paul Mauser, he revolutionized firearm design by introducing the first successful bolt-action system in 1871. The Mauser Model 1871 rifle featured a robust rotating bolt head that locked securely into the receiver, ensuring exceptional reliability and accuracy. This groundbreaking design quickly became a benchmark for future firearms development.

The success of the Model 1871 led to further refinements in bolt-action technology by Wilhelm Mauser. He introduced features such as a controlled-feed mechanism, which ensured smooth feeding and reliable extraction of cartridges under adverse conditions. These innovations significantly improved the overall functionality and performance of firearms.

Wilhelm's commitment to continuous improvement was evident in his later designs as well. The iconic Mauser Model 1898 rifle, often referred to as "the father of all modern bolt-action rifles," showcased his unwavering dedication to excellence. Its strong yet lightweight construction, superior ballistics, and smooth action set new standards for accuracy and reliability that are still admired today.

Another area where Wilhelm made lasting contributions was in cartridge development. He played a crucial role in designing several influential cartridges that became widely adopted around the world. The powerful 7x57mm (also known as .275 Rigby) cartridge developed by Wilhelm offered superior long-range performance while maintaining manageable recoil—a remarkable achievement at that time.

Moreover, Wilhelm's work laid the foundation for future advancements in firearm manufacturing techniques. Under his guidance, production methods were refined and standardized at Mauser Firearms, ensuring consistent quality and precision. The meticulous attention to detail in every aspect of production set a new benchmark for the industry, inspiring other manufacturers to follow suit.

Wilhelm Mauser's enduring impact on modern firearm design is undeniable. His pioneering bolt-action system, innovative features, and dedication to excellence have shaped the evolution of firearms for generations. Even today, many contemporary rifle designs owe their success to the principles established by Wilhelm over a century ago.

In recognition of his contributions, Wilhelm Mauser was honored with numerous awards and accolades during his lifetime. His legacy lives on through the continued success and popularity of Mauser Firearms, which remains an iconic brand associated with exceptional craftsmanship and performance.

Wilhelm Mauser's remarkable innovations continue to influence modern firearm design. His contributions in bolt-action systems, cartridge development, manufacturing techniques, and unwavering commitment to excellence have left an indelible mark on the industry. The enduring legacy of Wilhelm Mauser serves as an inspiration for engineers and designers alike who strive for perfection in their pursuit of firearm innovation.

***

In the vast realm of firearm history, few names stand as tall as Wilhelm Mauser. A visionary engineer and master craftsman, his innovations revolutionized the field of firearms and left an indelible mark on weapon design that continues to shape the industry to this day. As we celebrate his genius, it is essential to recognize the profound impact he had on the world of firearms.

Born in 1834 in Oberndorf am Neckar, Germany, Wilhelm Mauser began his journey towards greatness at a young age. Alongside his older brother Paul Mauser, he established a partnership that would become legendary in firearm history. Together, they founded Waffenfabrik Mauser, a company that would go on to produce some of the most iconic and influential firearms ever created. Wilhelm's brilliance lay not only in his technical expertise but also in his ability to innovate.

He developed numerous groundbreaking designs throughout his career that set new standards for firearm performance and reliability. One of these innovations was the introduction of smokeless powder ammunition—an advancement that dramatically improved both accuracy and safety. Perhaps Wilhelm Mauser's most significant contribution was the development of the Mauser bolt-action rifle system. This revolutionary design featured a rotating bolt head with multiple locking lugs, ensuring exceptional strength and durability while maintaining smooth operation.

This innovation became known as "the mauser action" and set a new benchmark for bolt-action rifles across the globe. The legacy of Wilhelm's work is particularly evident in one of his most iconic creations—the Mauser Model 98 rifle. Introduced in 1898, this rifle became one of the most popular military rifles ever produced. Its unmatched reliability, accuracy, and robustness made it highly sought after by armies around the world during both World Wars.

Beyond military applications, Wilhelm's designs also found success among civilian shooters and hunters. The Mauser Model 98's reputation for precision and performance made it a favorite among sportsmen worldwide. Even today, many enthusiasts consider the Mauser Model 98 as the pinnacle of bolt-action rifle design. Wilhelm Mauser's impact extended far beyond his own lifetime. His designs and innovations inspired generations of firearm engineers and manufacturers who sought to replicate his success.

Many of his concepts, such as the controlled-feed system, are still widely used in modern firearms. As we honor Wilhelm Mauser's legacy, it is essential to recognize not only his technical achievements but also the values he embodied. His dedication to quality craftsmanship and relentless pursuit of excellence set him apart as a true pioneer in firearm design. Wilhelm Mauser remains an enduring symbol of innovation and ingenuity, reminding us that with passion and perseverance, one can leave an indelible mark on history.

Celebrating Wilhelm Mauser means celebrating not only a legend but also the profound impact he had on firearm history. His genius innovations forever altered weapon design and continue to shape the industry today. Wilhelm's legacy lives on through his revolutionary bolt-action rifle system and timeless contributions to both military and civilian firearms.

Hiram Maxim
1840-1916

Throughout history, countless inventors have left their mark on society through their groundbreaking creations. One such innovator, whose name echoes through time, is Hiram Maxim. Known as the father of automatic weapons, Maxim revolutionized warfare with his invention of the machine gun. His brilliance and ingenuity paved the way for modern firearms and forever changed the course of history. Born on February 5, 1840, in Sangerville, Maine, Hiram Percy Maxim displayed an early aptitude for mechanics and engineering.

As a young boy growing up in a humble household, he would often tinker with various gadgets and machines in his spare time. This innate curiosity and passion for invention would eventually lead him down a path that would reshape military tactics and weaponry. Maxim's journey into the world of firearms began when he moved to England at the age of 41.

It was during this period that he truly embraced his calling as an inventor. In 1883, he successfully patented his first significant invention—a device known as "Maxim's Captive Flying Machine." Although this venture did not achieve commercial success, it showcased his inventive spirit and set him on a path toward greatness. However, it was not until several years later that Maxim would unveil his most significant creation—the machine gun.

In 1884, after extensive research and experimentation with various designs, he patented what would become known as the "Maxim Gun." This fully automatic weapon utilized recoil energy to reload itself continuously while firing multiple rounds per minute—an innovation that surpassed all previous firearm technology. The introduction of the machine gun marked a turning point in military history. Before its inception, battles were predominantly fought using single-shot rifles or manual repeating firearms.

The machine gun not only increased firepower exponentially but also introduced unprecedented levels of efficiency on the battlefield. The British Army quickly recognized the potential impact of Maxim's invention and adopted the Maxim Gun as its official machine gun in 1889. Its deployment during various colonial conflicts, such as the Anglo-Zulu War and the Boer War, showcased its devastating effectiveness against enemy forces.

The machine gun's ability to mow down adversaries with astonishing speed and accuracy forever changed military strategies and tactics. Maxim's contributions extended beyond his initial invention. He continued to refine and enhance his designs, resulting in lighter, more portable versions of the machine gun. His inventions were widely adopted by armies around the world, solidifying his place as one of history's most influential inventors.

Beyond his achievements in weaponry, Hiram Maxim was also an accomplished engineer and inventor in other fields. He developed numerous innovative devices like a steam pump regulator, a hair-curling iron, an automatic sprinkler system for fire protection, and even an amusement ride called "The Captive Flying Machine." His diverse range of inventions showcased his versatile mind and inventive prowess. Hiram Maxim's indelible mark on history cannot be overstated.

Through his relentless pursuit of innovation and unwavering dedication to engineering excellence, he forever transformed warfare with the introduction of the machine gun.

***

Hiram Maxim, born on February 5, 1840, in Sangerville, Maine, was an American-born inventor and engineer who made significant contributions to the field of mechanical engineering. His ingenuity and determination laid the foundation for many groundbreaking inventions that revolutionized various industries. Maxim's early life and influences played a crucial role in shaping his career and fueling his passion for innovation.

Growing up in rural Maine, Maxim displayed a natural curiosity and aptitude for mechanics from an early age. He was fascinated by machinery and spent countless hours tinkering with gadgets in his father's workshop. This hands-on experience allowed him to develop a deep understanding of mechanical principles even before receiving any formal education.

Maxim's father, Isaac Weston Maxim, was also an inventor and held numerous patents himself. He encouraged young Hiram's interests by providing him with books on mechanics and engineering principles. These resources became invaluable to the young inventor as he delved deeper into understanding the intricacies of machinery.

In 1854, at the age of 14, Maxim moved to Boston to work as an apprentice mechanic at a machine shop. This experience exposed him to advanced machinery and provided him with practical knowledge that would shape his future inventions. During this time, he honed his skills in metalworking, toolmaking, and precision machining.

In 1861, at the onset of the American Civil War, Maxim enlisted in the Union Army as a mechanic. Serving as a technical expert within various artillery units throughout the war further expanded his knowledge of weaponry systems. It was during this time that he began experimenting with improving existing firearms designs.

After leaving military service in 1865 following the end of the Civil War, Maxim moved to England where he continued pursuing his passion for inventing. The British industrial landscape provided him with access to resources not readily available elsewhere at that time.

Influenced by his experiences during the war, Maxim focused on developing more efficient and reliable firearms. In 1884, he patented the Maxim machine gun, which became one of his most significant inventions. This fully automatic weapon utilized recoil energy to reload ammunition, allowing for rapid fire with minimal manual effort. The Maxim machine gun would go on to revolutionize warfare and become the precursor to modern-day automatic weapons.

Maxim's success with the machine gun propelled him to international prominence as an inventor. He established the Maxim Gun Company in London in 1885 to manufacture and market his invention worldwide. This venture not only solidified his position as a leading figure in the arms industry but also provided him with financial stability.

Throughout his life, Hiram Maxim remained dedicated to innovation and continued inventing until his death on November

24, 1916. His early life experiences, including exposure to machinery at a young age, apprenticeship in Boston's machine shop, military service during the Civil War, and subsequent move to England all played integral roles in shaping his career trajectory and influencing his groundbreaking inventions.

Hiram Maxim's contributions continue to impact various industries and are a testament to his relentless pursuit of technological advancements.

***

In the late 19th century, warfare was on the brink of a significant transformation. The development of new technologies and inventions aimed to increase firepower and improve military strategies. Among these groundbreaking innovations, one name stands out: Hiram Maxim. His creation, the machine gun, would forever change the face of warfare and become an integral part of modern armies.

Hiram Percy Maxim was born on February 5, 1840, in Sangerville, Maine. From an early age, he displayed a fascination with machinery and engineering. As he grew older, his passion led him to pursue various projects and invent groundbreaking devices. However, it was his work on automatic firearms that would secure his place in history.

Maxim's journey towards developing the machine gun began in 1883 when he invented the first fully automatic portable machine gun known as the "Maxim Gun." Unlike its predecessors that required manual reloading after each shot, Maxim's design utilized recoil energy to automatically reload cartridges from a belt or magazine.

The key innovation behind Maxim's machine gun was its use of a toggle-joint mechanism that allowed for efficient extraction and ejection of spent cartridges while simultaneously loading fresh ammunition. This breakthrough ensured continuous fire without interruption or manual intervention—a feat never before achieved in firearms technology.

To further enhance its capabilities, Maxim incorporated a water-cooling system into his design. By introducing water into a jacket surrounding the barrel, he effectively minimized overheating issues during prolonged firing sessions. This feature increased reliability while maintaining a high rate of fire—a crucial advantage over other firearms at that time.

Maxim showcased his invention at numerous exhibitions across Europe and North America throughout the late 19th century. Its performance left military officials astounded by its potential impact on future battlefields. Recognizing its significance, several nations adopted Maxim's machine gun into their arsenals.

In 1889, Hiram Maxim founded the Maxim Gun Company, later renamed Vickers, Sons & Maxim. The company's primary focus was to manufacture and distribute his revolutionary firearm. Over time, improvements were made to the original design, resulting in more efficient and reliable models.

During the late 19th and early 20th centuries, Maxim's machine gun played a pivotal role in several conflicts worldwide. From colonial expeditions to major wars like World War I, armies armed with machine guns gained a significant advantage on the battlefield. The sheer firepower and sustained rate of fire offered by these weapons revolutionized military tactics and strategies.

Hiram Maxim's invention not only shaped warfare but also influenced subsequent developments in firearms technology. It paved the way for further advancements in automatic weaponry that would eventually lead to modern-day firearms being used by military forces around the globe.

Hiram Maxim's creation of the machine gun was a revolutionary breakthrough that forever changed the nature of warfare. His innovative design allowed for continuous fire without manual reloading and significantly increased firepower on the battlefield. With its incorporation into military arsenals worldwide, this invention laid the foundation for future developments in automatic weaponry and solidified its place as one of history's most impactful inventions.

***

The invention of the Maxim machine gun by Hiram Maxim in 1884 revolutionized warfare, fundamentally altering the dynamics and strategies employed on the battlefield. This remarkable weapon not only increased firepower but also provided a significant advantage to those who possessed it. The impact of the Maxim machine gun on warfare was profound, transforming the nature of conflicts and leaving an indelible mark on military history.

First and foremost, the Maxim machine gun dramatically enhanced firepower for those employing it. Unlike previous firearms that required manual reloading after every shot, this innovative weapon featured a self-contained mechanism capable of firing hundreds of rounds per minute. This unprecedented rate of fire gave an overwhelming advantage to armies equipped with these weapons, as they could unleash a continuous hailstorm of bullets upon their enemies.

Consequently, battles became deadlier than ever before, causing unprecedented casualties among opposing forces.

Furthermore, the Maxim machine gun altered traditional battlefield tactics. Prior to its introduction, armies relied heavily on massed infantry charges or cavalry assaults to break enemy lines. However, with its ability to mow down large numbers of troops rapidly and efficiently, the Maxim machine gun rendered such tactics obsolete and suicidal. Armies had to adapt quickly by finding new strategies that could counteract this formidable weapon's devastating effects.

Another significant impact was seen in trench warfare during World War I. The static nature of trench warfare led both sides to utilize machine guns extensively as defensive weapons. The Maxims were mounted in fortified positions along the trenches and effectively turned them into death traps for any advancing enemy soldiers. Attempting to cross "No Man's Land" under intense gunfire from Maxims proved nearly impossible for opposing forces without suffering substantial losses.

Moreover, the psychological impact caused by these deadly weapons cannot be underestimated. Soldiers facing a barrage from Maxims often experienced extreme fear and disorientation due to their destructive power and relentless firing capability. This psychological effect further demoralized enemy troops and contributed to the overall success of those wielding the Maxim machine guns.

The impact of the Maxim machine gun was not limited to its immediate effects on the battlefield. Its invention spurred a rapid arms race among nations, with each seeking to develop their own version or acquire this formidable weapon. The widespread adoption of machine guns eventually led to an imbalance in military power, creating a significant advantage for those who possessed them.

This shift in power dynamics greatly influenced geopolitical strategies, as nations sought to secure these weapons or develop countermeasures against them.

Hiram Maxim's invention of the Maxim machine gun revolutionized warfare in numerous ways. Its increased firepower, alteration of tactics, influence on trench warfare, psychological impact on soldiers, and subsequent arms race all contributed to its profound and lasting impact. The Maxim machine gun forever changed the face of warfare and left an enduring legacy as one of the most influential weapons in military history.

***

Hiram Maxim, a renowned American-born British inventor, is widely celebrated for his groundbreaking invention of the first practical and fully automatic machine gun. However, his contributions to weaponry extended far beyond this singular achievement. Throughout his career, Maxim continued to push the boundaries of firearms technology and made significant advancements that revolutionized the field of armaments.

Following the success of his machine gun invention, Maxim focused on refining and improving its design. He recognized the need for a more portable and efficient weapon that could be easily maneuvered on the battlefield. As a result, he developed the light machine gun, which was lighter in weight and offered greater mobility compared to its predecessors. This innovation enabled soldiers to carry more ammunition and provided them with increased firepower on the front lines.

Another notable contribution by Maxim was his development of smokeless powder cartridges. Traditionally, firearms utilized black powder as their propellant, which produced a thick cloud of smoke upon firing that obstructed visibility and revealed a shooter's position. Recognizing this limitation, Maxim sought to create a cleaner-burning propellant that would minimize smoke production while maintaining efficiency. His invention of smokeless powder revolutionized small arms technology by enhancing accuracy while simultaneously reducing detection risk during combat.

Maxim's ingenuity also extended beyond conventional firearms into artillery systems. He devised an innovative recoil mechanism for heavy guns that significantly improved their stability and accuracy during firing. Prior to this development, artillery pieces often experienced substantial recoil forces that not only disrupted the aim but also posed safety hazards for crews operating them. By incorporating an effective recoil system into these weapons, Maxim greatly enhanced their overall performance while ensuring safer operation.

Moreover, Hiram Maxim delved into inventing various forms of ordinance that expanded military capabilities even further. One such creation was the high-explosive shell designed specifically for use in modern cannons or howitzers. These shells were capable of inflicting significant damage upon impact, making them invaluable in both offensive and defensive operations. Additionally, Maxim invented the shrapnel shell, which contained small metal fragments that would scatter upon detonation, causing widespread casualties among enemy forces within its blast radius.

Maxim's relentless pursuit of innovation also led him to experiment with aviation weaponry. During World War I, he developed a synchronized machine gun system that allowed aircraft to fire through their propellers without damaging them. This breakthrough enabled pilots to engage in aerial combat more effectively and marked a pivotal advancement in the history of aerial warfare.

Hiram Maxim's continued innovations in weaponry not only transformed the field of armaments but also played a crucial role in shaping modern military strategies and tactics. His contributions paved the way for subsequent advancements in firearms technology and established new standards for firepower, mobility, and accuracy on the battlefield. Maxim's relentless pursuit of excellence undoubtedly left an indelible mark on the history of weapons development and forever changed the face of warfare as we know it today.

***

Hiram Maxim, a brilliant inventor and engineer, revolutionized firearms technology with his groundbreaking invention - the silencer. Born on February 5, 1840, in Sangerville, Maine, Maxim's relentless pursuit of innovation led him to establish the Maxim Silencer Company, forever changing the landscape of firearms.

Inspired by his father's passion for engineering and machinery, Maxim developed a keen interest in inventing from an early age. He began his career as an electrical engineer and quickly gained recognition for his expertise in this field. However, it was his fascination with firearms that would ultimately propel him to global fame.

In the late 19th century, firearms were loud and cumbersome devices that often revealed one's position on the battlefield. Recognizing this significant drawback, Maxim set out to develop a solution that would enhance stealth and accuracy while reducing noise levels. After years of research and experimentation, he unveiled his crowning achievement - the first practical firearm silencer.

Maxim's silencer incorporated ingenious engineering principles that allowed for effective noise reduction without sacrificing firepower or reliability. Unlike previous attempts at creating suppressors that resulted in decreased muzzle velocity or caused excessive backpressure, leading to malfunctions,

Maxim's design efficiently dissipated gas pressure while maintaining optimal performance.

The success of his invention prompted Maxim to establish the Maxim Silencer Company in 1902. Located in Hartford, Connecticut, this new venture aimed to meet the growing demand for firearm suppressors across military forces worldwide. The company quickly gained recognition as a leading provider of cutting-edge firearm technologies.

Maxim's silencers found immediate application within military circles where stealth was paramount during covert operations and special missions. The British Army became one of their earliest clients when they equipped their troops with silenced pistols during World War I. This adoption marked a turning point in modern warfare tactics as it allowed soldiers to engage enemies discreetly, providing a significant advantage on the battlefield.

The Maxim Silencer Company's reputation continued to soar as their products found use in law enforcement agencies and civilian markets alike. The demand for suppressors expanded beyond military applications, with hunters, competitive shooters, and gun enthusiasts recognizing the benefits of reduced noise and recoil.

Hiram Maxim's contributions to firearms technology extended far beyond the silencer. His relentless pursuit of innovation led to numerous other inventions, including the first portable machine gun - the Maxim Gun. This invention further solidified his position as a pioneer in the field of firearms engineering.

The formation of the Maxim Silencer Company marked a watershed moment in firearms technology. Hiram Maxim's revolutionary invention forever changed how we perceive and utilize firearms. His silencers not only enhanced stealth and accuracy but also paved the way for future advancements in suppressor technology that continue to shape modern warfare strategies and civilian shooting sports today.

***

Hiram Maxim, an inventive genius and prolific inventor of the late 19th and early 20th centuries, left an indelible mark on modern warfare. His innovative creations revolutionized the battlefield, forever changing the nature of conflicts and shaping military strategies for decades to come. Maxim's contributions were not limited to a single invention; his legacy encompasses a range of groundbreaking advancements that continue to influence military technology today.

One of Maxim's most notable contributions was the invention of the first practical fully automatic machine gun. In 1884, he patented the Maxim Gun, a weapon capable of sustained fire without manual reloading. This innovation eliminated the need for large numbers of soldiers to engage in battle, as one operator could now effectively suppress enemy forces. The maxim gun provided superior firepower and played a significant role in shaping modern infantry tactics.

Furthermore, Maxim's inventive prowess extended beyond firearms. He also made substantial advancements in explosives technology by inventing smokeless powder. Traditionally, black powder had been used as a propellant for firearms but produced copious amounts of smoke, obscuring visibility on the battlefield. Maxim's smokeless powder was a game-changer as it significantly reduced muzzle flash and smoke emission while maintaining projectile velocity and accuracy.

This development allowed armies to engage in combat with increased efficiency and precision.

Maxim's contributions to modern warfare were not solely focused on weaponry; he also revolutionized transportation during times of conflict. Recognizing the potential for armored vehicles on battlefields, he designed an early prototype known as "Maxim's horseless carriage." Although this prototype did not see widespread use during his lifetime, it laid the foundation for future developments in armored vehicles that would play pivotal roles in both World Wars.

Another aspect of warfare that Maxim influenced was aerial combat through his development of aircraft machine guns. In collaboration with aviation pioneers like Sir Hiram S. Maxim and his son, Hiram Percy Maxim, he created lightweight and efficient machine guns specifically designed for aircraft use. These weapons became instrumental in aerial warfare, allowing pilots to engage enemy aircraft with precision and accuracy.

The legacy of Hiram Maxim's contributions to modern warfare is undeniable. His innovations not only changed the way wars were fought but also influenced the strategies employed by military forces worldwide. The fully automatic machine gun, smokeless powder, armored vehicles, and aircraft machine guns are just a few examples of his groundbreaking inventions that continue to shape modern military technology.

Maxim's impact on warfare extends far beyond his own time; his inventions laid the foundation for subsequent advancements in weaponry and tactics. Today, we see echoes of Maxim's innovations in the advanced firearms used by soldiers on the battlefield, as well as in the armored vehicles and aircraft employed by armed forces worldwide. Hiram Maxim's legacy stands as a testament to human ingenuity and its ability to reshape the world through technological advancements that forever change the course of history.

***

When we hear the name Hiram Maxim, many of us immediately think of the famous machine gun that bears his name. However, what often goes unnoticed are the numerous other inventions and innovations that this remarkable engineer contributed to society. From steam-powered airplanes to electrical devices, Hiram Maxim's lesser-known inventions showcase his brilliance and his wide-ranging interests beyond weaponry.

One of Maxim's most fascinating yet forgotten innovations was his work on steam-powered flight. While the Wright brothers are often credited with inventing the first successful airplane, it is worth noting that Maxim experimented with steam-powered flight well before their iconic flight at Kitty Hawk in 1903. In fact, as early as 1894, Maxim constructed a gigantic aircraft called "Aerial Steam Carriage," which had a wingspan of nearly 100 feet.

Although this ambitious project did not achieve sustained flight, it laid the groundwork for future aviation pioneers.

Another forgotten innovation by Hiram Maxim lies in the field of electrical engineering. At a time when electricity was still in its infancy, Maxim invented several devices that greatly impacted daily life. One such invention was an improved carbon filament lamp that increased the efficiency and longevity of electric light bulbs. His advancements in lighting technology paved the way for widespread adoption and contributed to the rapid electrification of cities around the world.

Additionally, Maxim made significant contributions to transportation systems through his innovative designs for steam-powered engines and locomotives. He developed a revolutionary boiler design known as "Maxim's Boiler," which improved efficiency and safety in steam engines by reducing pressure fluctuations. This innovation not only enhanced locomotive performance but also had wide-ranging applications across various industries reliant on steam power.

Maxim's inventive mind also extended into other areas such as heating systems and firefighting equipment. He developed a novel heating system known as "Maxim Radiant Heat," which utilized hot water and steam to efficiently warm buildings. This system revolutionized heating technology and found widespread use in homes, offices, and factories. Furthermore, his inventions in firefighting equipment, such as the "Maxim Automatic Fire Sprinkler," greatly improved fire safety measures, ultimately saving countless lives.

While the Maxim machine gun remains his most enduring legacy, it is essential to recognize Hiram Maxim's lesser-known inventions that have had a lasting impact on society. From his pioneering work in aviation to advancements in electrical engineering and transportation systems, Maxim's innovative spirit touched various aspects of human life. Through these forgotten innovations, we gain a deeper appreciation for the breadth of his contributions and the extent of his genius.

Hiram Maxim's brilliance extended far beyond the realm of weaponry. His lesser-known inventions stand as a testament to his inventive mind and demonstrate his significant impact on diverse fields such as aviation, electrical engineering, transportation systems, heating technology, and fire safety. By exploring these forgotten innovations by Hiram Maxim, we can truly appreciate the breadth of his contributions to society and recognize him as more than just the inventor of a famous machine gun.

***

Hiram Maxim, a remarkable inventor and engineer, made groundbreaking contributions to the field of weaponry during the late 19th and early 20th centuries. His innovative designs revolutionized firearms technology, forever changing the landscape of warfare. Maxim's brilliance did not go unnoticed, as he received numerous recognitions and awards throughout his illustrious career.

One of the most prestigious accolades bestowed upon Hiram Maxim was the Royal Victorian Order (RVO), which he received in 1901. The RVO is a dynastic order of knighthood established by Queen Victoria in 1896 to recognize distinguished personal service to the British monarch. This honor highlighted Maxim's extraordinary contributions to military science and his dedication to advancing weaponry for the British Empire.

Maxim's incredible inventions also earned him international recognition. In 1888, he was awarded the Grand Prix at the Exposition Universelle in Paris for his revolutionary recoil-operated machine gun—the Maxim Gun. This recognition not only solidified his status as an exceptional inventor but also brought him global acclaim as one of history's greatest weapon designers.

In addition to individual awards, Hiram Maxim's inventions received widespread acknowledgment from various military institutions across Europe and North America. The British Army adopted several versions of his machine guns, including the iconic Vickers machine gun that served as a staple for British forces throughout World War I. These weapons garnered immense respect from soldiers on all fronts due to their exceptional reliability and firepower.

Maxim's contributions were not limited to just firearms; he also played a crucial role in developing artillery technology. His work on designing smokeless powder—a significant advancement over traditional black powder—revolutionized artillery shells' effectiveness and range. Recognizing this achievement, Maxim was honored with multiple awards by several prestigious engineering societies for his invaluable contributions towards modernizing artillery systems.

Moreover, Hiram Maxim's accomplishments were not confined to the realm of military applications. His inventive spirit extended to civilian life as well. In 1884, he patented the first practical and commercially viable electric lightbulb. Although Thomas Edison is often credited with this invention, it was Maxim who successfully demonstrated the lightbulb's feasibility on a large scale, earning him recognition as a pioneer in electrical engineering.

The legacy of Hiram Maxim's achievements continues to be celebrated even today. His inventions shaped the course of warfare and propelled technological advancements across various fields. By recognizing his contributions through awards and honors, society acknowledges the profound impact he had on military science, engineering, and even everyday life.

Hiram Maxim's remarkable career as an inventor and engineer garnered him numerous recognitions and awards throughout his life. From receiving knighthood through the Royal Victorian Order to being awarded international accolades for his groundbreaking machine gun designs, Maxim's contributions were widely celebrated both within military circles and among engineering societies. His legacy lives on as his inventions continue to shape modern warfare, while his pioneering work in electrical engineering remains influential in our everyday lives.

***

Hiram Maxim, an influential inventor and engineer, left an indelible mark on the world through his groundbreaking innovations that continue to impact various aspects of our lives. From his revolutionary machine gun to his advancements in electricity and aviation, Maxim's contributions have shaped modern technology. Today, we celebrate the life and achievements of this remarkable pioneer.

Born on February 5, 1840, in Sangerville, Maine, Hiram Maxim demonstrated an early aptitude for engineering. His inventive spirit led him to create numerous inventions throughout his lifetime. However, it was his invention of the world's first portable fully automatic machine gun that revolutionized warfare and earned him lasting recognition.

Maxim's machine gun design was patented in 1883 and quickly gained popularity due to its reliability and efficiency on the battlefield. Unlike previous firearms that required manual reloading after each shot, Maxim's invention utilized recoil energy to reload automatically. This breakthrough innovation paved the way for modern automatic weapons that are still used today.

Beyond his achievements in weaponry, Maxim also made significant contributions to other fields such as electricity and aviation. In collaboration with Thomas Edison, he developed a new type of electric lamp that significantly improved efficiency compared to existing models at the time. His work also extended into aviation when he designed a large biplane known as "The Captive Flying Machine." Although it never achieved successful flight due to its immense size and weight limitations at the time, this aircraft laid the groundwork for future advancements in aviation technology.

Maxim's brilliance extended beyond inventing; he was also a prolific writer who shared his knowledge with others through numerous publications. His book titled "Artificial and Natural Flight" explored theories behind human flight while providing insights into aerodynamics—a subject that fascinated him until his final days.

Throughout his life, Hiram Maxim received several accolades for his remarkable contributions to science and technology. In 1901, he was knighted by Queen Victoria for his services to the British Empire. Maxim's achievements continue to be celebrated through various honors and exhibitions dedicated to his work, ensuring that future generations recognize the impact of his inventions.

Today, we remember Hiram Maxim as a pioneer whose technological advancements have shaped our world. His machine gun revolutionized warfare, while his contributions to electricity and aviation laid the groundwork for future innovations in these fields. Maxim's legacy serves as an inspiration for inventors and engineers worldwide, reminding us of the power of human ingenuity and its ability to transform society.

As we celebrate the life of Hiram Maxim, let us be reminded of his unwavering passion for innovation and commitment to pushing the boundaries of what is possible. His legacy continues to inspire generations as we strive towards new frontiers in science and technology, building upon the foundations laid by this extraordinary pioneer.

Georg Luger
1849-1923

In the realm of firearms, certain names stand out as pioneers and masterminds behind legendary designs. One such name is Georg Luger, the brilliant Austrian firearms designer who left an indelible mark on the history of weaponry with his iconic Luger pistol. Revered for its sleek aesthetics, innovative engineering, and exceptional performance, the Luger pistol became synonymous with reliability and sophistication.

This subtopic delves into the life and legacy of Georg Luger, exploring his journey as a firearms designer and the enduring impact of his masterpiece. Georg Johann Heinrich Luger was born on March 6, 1849, in Steinach am Brenner, Austria. From an early age, he showed a keen interest in mechanics and engineering. After completing his education at a technical school in Vienna, he embarked on a career that would shape the future of firearms design.

Starting as an apprentice at various firearm manufacturing companies across Europe, Luger honed his skills before eventually joining Ludwig Loewe & Company in Berlin. It was during his time at Ludwig Loewe & Company that Luger's ingenious mind began to flourish. In collaboration with Hugo Borchardt—an accomplished German engineer—Luger developed what would become one of the most iconic handgun designs ever conceived: the toggle-locked blowback action system.

This revolutionary mechanism allowed for increased accuracy by reducing recoil while maintaining optimal reliability—a hallmark feature of the subsequent Lugers. In 1900, after years of relentless refinement and innovation under various iterations, Georg Luger unveiled what would become his magnum opus—the P08 Parabellum Pistol or simply known as the "Luger." Its distinctive shape, with its characteristic grip angle and sleek lines, made it instantly recognizable among gun enthusiasts worldwide.

The success of the Luger pistol extended beyond its impressive aesthetics; it excelled in terms of performance as well. Chambered in the powerful 9x19mm Parabellum cartridge, the Luger offered superior accuracy and a groundbreaking toggle-lock system that ensured reliable feeding and ejection. Its unique toggle action, combined with an innovative recoil system, made the Luger exceptionally controllable during rapid fire—an attribute highly valued by military forces.

The Luger pistol gained immense popularity among military personnel and law enforcement agencies across Europe. It was adopted as the standard sidearm for various armed forces, including Germany, Switzerland, Portugal, and Finland. Its use continued well into the 20th century until it was gradually replaced by more modern designs. Today, the legacy of Georg Luger lives on through his iconic pistol.

Collectors worldwide cherish Lugers not only for their historical significance but also for their craftsmanship and engineering excellence. These firearms have become highly sought-after pieces that exemplify Luger's vision of blending form with function. Georg Luger's contributions to firearms design cannot be overstated. His innovative mind and meticulous attention to detail created a masterpiece in the form of the iconic Luger pistol—a symbol of elegance, reliability, and precision engineering.

***

Georg Johann Luger, renowned as the mastermind behind the iconic Luger pistol, was born on March 6, 1849, in Steinach am Brenner, Tyrol. His early life was marked by a deep fascination with mechanical engineering and a relentless pursuit of knowledge that would eventually shape his illustrious career.

Growing up in a small village nestled in the Austrian Alps, Luger exhibited an innate curiosity for understanding how things worked.

He spent countless hours tinkering with various gadgets and machinery he found around his home, much to the chagrin of his parents. Recognizing their son's exceptional aptitude for mechanics at an early age, they nurtured his passion by allowing him to explore and experiment freely. After completing his basic education in Steinach am Brenner, Luger enrolled at the Polytechnic Institute in Vienna.

The institute offered a comprehensive curriculum that encompassed mathematics, physics, chemistry, and engineering disciplines – all of which proved instrumental in shaping Luger's future career path. During his time at the Polytechnic Institute, Luger developed a particular interest in firearms technology. He avidly studied existing designs and theories surrounding firearms mechanisms while also delving into related subjects such as ballistics and metallurgy.

His zeal for understanding every aspect of firearm design led him to spend numerous hours conducting hands-on experiments at shooting ranges or collaborating with local gunsmiths. Upon graduating from the institute in 1871 with top honors and earning a degree in mechanical engineering specializing in firearms technology, Luger sought employment opportunities that would allow him to apply his newfound knowledge effectively.

He joined Ludwig Loewe & Company (Loewe), one of Germany's leading arms manufacturers based in Berlin. At Loewe's engineering division under Hugo Borchardt's supervision – an accomplished firearm designer himself – Luger honed his skills and further expanded his knowledge of firearms design. Borchardt's mentorship played a pivotal role in Luger's development as an engineer, fostering an environment that encouraged innovation and creativity.

Luger's breakthrough came in 1898 when he patented the design for what would become his most significant contribution to the firearms industry – the Luger pistol. This semi-automatic handgun featured a toggle-lock mechanism, which offered superior reliability, accuracy, and ease of use compared to existing designs. Its sleek appearance, ergonomic grip, and innovative features made it highly coveted by military forces and civilians alike.

Throughout his career, Luger continued to refine his designs while also collaborating with prominent arms manufacturers such as Deutsche Waffen und Munitionsfabriken (DWM) to bring his creations to life. His relentless pursuit of perfection led him to create several variations of the iconic Luger pistol that further solidified its reputation as one of the most advanced handguns in history. Georg Luger's early life experiences and unwavering dedication to mechanical engineering laid the foundation for a remarkable career that forever changed firearms technology.

***

Georg Luger, a name synonymous with firearms history, is widely regarded as one of the most influential firearm designers of all time. His groundbreaking work in the late 19th and early 20th centuries revolutionized handgun design and led to the creation of one of the most iconic pistols ever made – the Luger pistol. Luger's contribution to handgun innovations cannot be overstated, as his designs set new standards for reliability, accuracy, and functionality that are still revered today.

Luger began his career as an engineer in Austria-Hungary's Royal Arsenal at Brno, where he gained invaluable experience in firearms design. It was during this time that he began exploring ideas for improving existing handguns. In 1898, after years of experimentation and refinement, Luger introduced his masterpiece – the Parabellum Pistol or what later became known as the Luger pistol.

What set Luger's design apart from its contemporaries was its unique toggle-lock action mechanism. This innovative system utilized a jointed arm that locked the barrel and breech together during firing. The toggle-lock allowed for faster cycling rates and reduced recoil compared to other semiautomatic pistols at that time. This breakthrough not only improved accuracy but also made it possible for shooters to fire rapidly without compromising reliability.

Another significant contribution made by Luger was his introduction of a shoulder stock attachment for his pistol. By attaching a wooden stock to the grip frame of the pistol, it transformed into a compact carbine-like weapon known as the "Artillery" model. This innovation greatly increased stability and accuracy when firing at longer ranges while maintaining portability.

Luger's commitment to excellence extended beyond mechanical innovations; he also prioritized ergonomics in his designs. He recognized that comfortable handling was vital for accurate shooting and designed his pistols with an ergonomic grip angle that fit naturally in the shooter's hand. This attention to detail set a new standard for handgun design and greatly influenced future generations of pistols.

The Luger pistol's success was not limited to its mechanical superiority but also owed much to Luger's marketing prowess. He tirelessly promoted his creation, ensuring its widespread adoption by numerous military forces around the world. The German Army, in particular, embraced the Luger pistol as their standard-issue sidearm during both World Wars, further solidifying its legendary status.

Georg Luger's contribution to handgun innovations cannot be underestimated. His pioneering work in firearm design and engineering gave birth to a legend that remains unrivaled even today. The toggle-lock action mechanism, shoulder stock attachment, and ergonomic grip angle introduced by Luger set new benchmarks for performance, reliability, and user comfort that have endured over a century later.

Although Georg Luger passed away in 1923, his name lives on through his revolutionary designs and the iconic pistol that bears his name. His visionary contributions continue to inspire firearm enthusiasts and designers worldwide as they strive to build upon his legacy of excellence in handgun innovation.

****

When discussing iconic firearms, one cannot overlook the Luger pistol, a masterpiece in design and engineering. Crafted by the brilliant mind of Georg Luger, this firearm not only revolutionized the world of handguns but also became a symbol of German engineering excellence. The Luger pistol's unique features and innovative design elements set it apart from its contemporaries, making it an enduring icon.

One of the most striking features of the Luger pistol is its toggle-lock mechanism. Unlike other pistols that rely on a traditional slide to load and eject rounds, Lugers employ a toggle-action system that provides enhanced reliability and accuracy. This mechanism consists of two interconnected toggles that lock into place during firing, ensuring optimal alignment between barrel and chamber for each shot.

The smooth and precise operation of this system contributes to the exceptional accuracy for which Lugers are renowned.

Another remarkable aspect of Lugers is their grip design. Georg Luger recognized early on that ergonomics played a crucial role in firearm performance, leading him to develop an innovative grip angle for his pistol. By angling the grip slightly downwards, he achieved better recoil management and improved control during rapid fire. This ergonomic advantage made Lugers highly sought-after by military personnel and law enforcement agencies around the world.

The aesthetics of the Luger pistol also contribute to its timeless appeal. With its sleek lines, well-defined edges, and distinctive shape, this firearm exudes elegance in both form and function. The attention to detail in every aspect of its design reflects Georg Luger's commitment to creating not just a reliable weapon but also a work of art.

Additionally, Lugers were among the first semi-automatic pistols to incorporate a detachable magazine—a groundbreaking feature at that time. This innovation allowed for quick reloading during combat situations without requiring individual rounds to be manually loaded into the chamber. The detachable magazine system significantly increased the pistol's firepower and efficiency, making it a formidable weapon on the battlefield.

Furthermore, the Luger's caliber options added to its versatility. Initially chambered in 7.65mm Parabellum, Lugers were later produced in 9mm Parabellum—a cartridge that became widely adopted by militaries worldwide. The ability to interchange barrels and magazines allowed users to adapt their Lugers to different ammunition types, further enhancing their practicality.

Georg Luger's vision of designing a firearm that combined exceptional performance with exquisite aesthetics resulted in the creation of the iconic Luger pistol. Its toggle-lock mechanism, ergonomic grip angle, attention to detail, detachable magazine system, and caliber options set new standards for handgun design. Today, over a century after its inception, the Luger pistol remains an enduring symbol of excellence and continues to captivate firearm enthusiasts around the world.

***

The semi-automatic Luger pistol, designed by Georg Luger, is widely recognized as one of the most iconic firearms in history. Its innovative design and reliable functionality revolutionized the world of handguns. However, the development process of this legendary weapon was not without its challenges and complexities.

The journey from concept to reality began in the late 19th century when Georg Luger, a talented Austrian engineer, set out to create a firearm that would surpass all others in terms of reliability and performance. Luger's vision was to design a semi-automatic pistol that would combine high firepower with simplicity and ease of use.

Luger started by studying existing firearms, analyzing their strengths and weaknesses. He sought inspiration from various sources, including earlier semi-automatic designs like the Borchardt C-93 pistol. This extensive research allowed him to identify areas for improvement and set clear goals for his own creation.

In 1898, after years of meticulous planning and experimentation, Luger unveiled his masterpiece – the Parabellum pistol, which later became known as the Luger pistol. The key feature that distinguished this firearm from others was its toggle-lock mechanism. This unique system utilized a jointed arm that connected the barrel and breechblock assembly to the frame. As recoil forces were generated upon firing a round, this toggle-lock system would absorb some of these forces before unlocking and allowing for automatic ejection of spent cartridges.

To ensure utmost reliability and performance, extensive testing was conducted on various prototypes before finalizing the design for mass production. These tests involved firing thousands of rounds under different conditions such as extreme temperatures or adverse environments. Each prototype underwent rigorous scrutiny for any potential flaws or malfunctions.

Once satisfied with its performance during testing phases, manufacturing began on a larger scale. Skilled craftsmen meticulously machined each component to exacting specifications before assembling them into a fully functional firearm. This attention to detail and precision craftsmanship ensured that every Luger pistol met the highest standards of quality.

The Luger pistol quickly gained popularity among military forces and civilians alike. Its sleek design, exceptional accuracy, and ability to accommodate various calibers made it a sought-after weapon. Over the years, the design evolved with improvements such as reinforced toggle locks and adjustable sights, further enhancing its performance.

Today, the Luger pistol remains an iconic symbol of innovation in firearm design. Its legacy lives on through countless replicas and its influence can be seen in subsequent semi-automatic handguns. Georg Luger's dedication to creating a reliable and efficient firearm resulted in a timeless masterpiece that continues to captivate gun enthusiasts around the world.

The development process of the semi-automatic Luger pistol was a journey fueled by passion, research, and meticulous testing. From its humble beginnings as an idea in Georg Luger's mind to becoming one of the most iconic firearms ever created, this legendary weapon stands as a testament to human ingenuity and engineering excellence.

***

In the realm of firearms, few names resonate with as much historical significance and engineering prowess as Georg Luger. The iconic Luger pistol, also known as the Parabellum, became a symbol of German engineering excellence during the late 19th and early 20th centuries. Its rise to prominence was not only marked by technical innovations but also by its association with military might and precision craftsmanship.

Georg Luger, born in Austria in 1849, was a talented engineer who dedicated his life to perfecting firearms design. After joining the renowned firearms manufacturer Deutsche Waffen- und Munitionsfabriken (DWM) in Germany in 1898, Luger's genius truly began to shine. It was during this time that he developed what would become his most famous creation—the self-loading semiautomatic pistol known as the Luger.

The Luger pistol quickly gained popularity due to its innovative design features and exceptional reliability. One of its defining characteristics was its toggle-lock mechanism, which provided a unique method for locking and unlocking the breechblock during firing cycles. This mechanism allowed for increased accuracy and reduced recoil, setting it apart from other contemporary pistols on the market. Moreover, the Luger's sleek aesthetics played an essential role in its rapid rise to prominence.

Its distinctive shape featuring a tapered barrel, ergonomic grip angle, and elegant lines made it instantly recognizable among gun enthusiasts worldwide. The attention to detail extended to every aspect of production—each component meticulously crafted with precision engineering techniques that were unparalleled at that time. The increasing popularity of the Luger pistol coincided with Germany's rising military power in the early 20th century.

As tensions escalated across Europe leading up to World War I, Germany sought both technological superiority on the battlefield and national prestige through their armaments industry. The adoption of Lugers as standard-issue sidearms for the German military further solidified its reputation as a symbol of German engineering excellence. The Luger's association with military might only grew stronger during World War II when it became the pistol of choice for high-ranking officers, including Adolf Hitler himself.

Its sleek design and association with power and authority further contributed to its iconic status. Even after the war, the Luger pistol continued to capture the imagination of firearms enthusiasts worldwide. Its place in popular culture was cemented through appearances in countless movies, books, and video games, perpetuating its image as a symbol of precision engineering and German craftsmanship. Today, owning an original Luger is considered a prized possession among collectors.

The scarcity and historical significance of these pistols have driven their prices to exorbitant levels at auctions. The legacy of Georg Luger lives on through his masterpiece—the Luger pistol—an enduring symbol of German engineering prowess that continues to captivate gun enthusiasts and historians alike. Georg Luger's relentless pursuit of firearm perfection gave birth to one of the most iconic pistols in history—the Luger.

***

When discussing iconic firearms from World War I, the Luger pistol undoubtedly takes center stage. Designed by Georg Luger, a brilliant Austrian engineer, this weapon played a significant role in shaping the course of the war. With its distinctive design and exceptional reliability, the Luger pistol became a symbol of power and precision on the battlefield.

Introduced in 1900, the Luger pistol quickly gained popularity among military forces worldwide due to its innovative features. Its semi-automatic operation allowed for rapid firing without manually reloading after every shot, giving soldiers a significant advantage over their adversaries armed with slower and less efficient firearms. Additionally, its toggle-lock mechanism ensured excellent accuracy by reducing recoil and improving stability during firing.

The Luger's design was so impressive that several countries adopted it as their standard-issue sidearm during World War I. Germany, in particular, heavily relied on this weapon throughout the conflict. German soldiers equipped with Lugers had an edge over their opponents due to its superior range and stopping power compared to other pistols of that era.

One key aspect that made the Luger pistol indispensable in trench warfare was its adaptability for use with accessories such as detachable shoulder stocks or snail drum magazines. These accessories transformed it into a compact carbine-like weapon capable of engaging targets at longer distances effectively. This versatility allowed soldiers to swiftly transition from close quarters combat to engaging enemies further away without needing to switch weapons.

Furthermore, ammunition compatibility played a crucial role in establishing the dominance of the Luger pistol on European battlefields. The 9mm Parabellum cartridge used by Lugers was widely available among various allied nations during World War I. This compatibility facilitated ammunition sharing between soldiers from different countries and increased overall operational efficiency.

The iconic look of the Luger pistol also contributed to its prominence on battlefields worldwide. Its sleek profile and distinctive shape made it instantly recognizable, instilling fear in the hearts of enemy combatants. The Luger's reputation as a symbol of German military might grew as the war progressed.

However, despite its undeniable success and popularity, the Luger pistol did have some drawbacks. Its complex design made it expensive and time-consuming to manufacture, which limited its availability on the front lines. Additionally, field maintenance was often challenging due to its intricate parts and precise tolerances. Nevertheless, these limitations did not diminish the Luger pistol's impact on World War I.

Georg Luger's innovative design revolutionized pistol technology during World War I. The Luger pistol's semi-automatic operation, adaptability with accessories, ammunition compatibility, and iconic appearance made it an invaluable weapon for soldiers on both sides of the conflict. Its widespread adoption by various nations is a testament to its effectiveness and enduring legacy as one of the most iconic firearms in history.

***

Georg Luger, a visionary firearm designer of the late 19th and early 20th centuries, left an indelible mark on the world of handguns with his most renowned creation—the iconic Luger pistol. This subtopic delves into the enduring influence of Georg

Luger's invention, exploring its impact on modern handguns and its significance within the firearms industry.

One cannot overlook the lasting legacy of the Luger pistol in terms of design innovation. Introduced in 1898, it featured several groundbreaking elements that revolutionized handgun engineering. The most notable was its toggle-lock mechanism—an ingenious design that enhanced reliability and accuracy. This innovative system provided a strong mechanical advantage, allowing for improved control over recoil forces compared to other contemporary designs.

Even today, many modern handguns incorporate similar locking mechanisms inspired by Georg Luger's original concept.

The Luger pistol also introduced a staggered magazine design that increased ammunition capacity without significantly increasing overall size or weight. This advancement set a precedent for subsequent handgun designs, as manufacturers recognized the benefits of maximizing firepower while maintaining portability—a principle still followed in modern firearms development.

Georg Luger's creation also significantly influenced ergonomics and user experience in handgun design. The grip angle and shape of the Luger pistol were carefully crafted to fit comfortably in hand, improving handling and control during shooting. These ergonomic considerations set new standards for future handguns and continue to shape contemporary designs aimed at optimizing user comfort and reducing fatigue.

Furthermore, Georg Luger's use of advanced manufacturing techniques was ahead of its time. He employed precision machining methods to ensure superior fit and finish in his pistols—a practice that has become standard in modern firearms production. Today's manufacturers continue to embrace such techniques to achieve high-quality results while minimizing manufacturing tolerances.

Beyond these technical advancements, Georg Luger's creation holds cultural significance within the firearms industry. The Luger pistol's distinctive aesthetics, with its sleek lines and iconic toggle-lock silhouette, have become synonymous with excellence in handgun design. Its timeless appeal has made it a sought-after collector's item and a symbol of craftsmanship.

Moreover, the Luger pistol's influence extends to popular culture as well. Its appearances in countless films, novels, and video games have cemented its status as an icon of firepower and elegance. This cultural impact has further solidified Georg Luger's legacy as a mastermind whose creation transcends the realm of firearms enthusiasts.

Georg Luger's invention—the Luger pistol—has left an enduring influence on modern handguns that extends far beyond its initial introduction over a century ago. Its innovative design elements continue to shape firearm engineering practices today, while its ergonomic considerations and manufacturing techniques have set new standards in user experience and production quality. Moreover, the cultural significance attached to this iconic handgun ensures that Georg Luger's legacy will persist for generations to come in both the firearms industry and popular culture.

***

In the world of firearms collecting, few weapons have captivated enthusiasts as much as the iconic Luger pistol. Developed by Georg Luger, a visionary firearms designer in the early 20th century, this masterpiece of engineering and aesthetics has become a true collector's favorite. Even a century after its creation, vintage Luger pistols continue to hold an irresistible allure for gun aficionados around the globe.

One of the primary reasons for the enduring fascination with vintage Lugers is their unmatched craftsmanship. These pistols were meticulously handcrafted using high-quality materials, resulting in unrivaled attention to detail and precision. From their sleek lines to their intricate engravings, every element of a vintage Luger reflects the dedication and skill of its maker. Collectors appreciate these fine details and recognize them as hallmarks of exceptional craftsmanship.

Moreover, Lugers are not just beautiful objects; they also possess incredible historical significance. Originally designed for military use, these pistols played a crucial role in World War I and World War II. Many collectors are drawn to Lugers because they symbolize an important era in global history. Owning one allows them to connect with those who used these firearms on battlefields long ago.

The scarcity of vintage Lugers further adds to their desirability among collectors. As time passes, finding well-preserved examples becomes increasingly challenging, making them highly sought after by enthusiasts. The limited supply drives up prices at auctions and makes owning such a piece both prestigious and exclusive.

Collectors also appreciate that each vintage Luger tells its own unique story through various markings or serial numbers on its frame or barrel. These markings often indicate which factory produced the pistol or provide insights into its production year or military issue details—making each firearm distinct from others within a collection.

Another aspect that adds allure to collecting Lugers is their mechanical ingenuity. Georg Luger's design was groundbreaking for its time, featuring a toggle-lock action and an innovative cartridge feeding mechanism. This mechanical complexity appeals to collectors who admire the technical achievements of firearms engineering. Owning a vintage Luger allows them to marvel at the innovation that went into creating such a remarkable weapon.

Finally, the community of Luger collectors adds an extra layer of fascination. These enthusiasts form a tight-knit group that shares their passion for Lugers through forums, gatherings, and exhibitions. The opportunity to connect with fellow collectors and learn from their expertise enriches the overall experience of collecting Lugers.

vintage Luger pistols continue to enthrall collectors worldwide due to their exceptional craftsmanship, historical significance, scarcity, unique markings, mechanical ingenuity, and vibrant community. These factors combine to create an enduring fascination with these iconic firearms. As long as there are collectors who appreciate fine craftsmanship and are captivated by historical artifacts, the allure of the vintage Luger pistol will remain undiminished in the years to come.

***

Georg Luger, the mastermind behind the iconic Luger pistol, has left an enduring legacy in the world of firearms technology. His innovative designs and contributions have revolutionized the industry, making him a legendary figure among gun enthusiasts and collectors.

Luger's most significant contribution was undoubtedly his creation of the Luger pistol itself. Introduced in 1900, it quickly gained popularity for its revolutionary toggle-lock system and superior ergonomics. The Luger pistol set new standards for reliability, accuracy, and ease of use. Its distinctive design with sleek lines and a well-balanced grip made it a coveted weapon both on and off the battlefield.

The toggle-lock mechanism invented by Luger was a breakthrough in firearms technology. It provided exceptional stability during firing by locking the barrel to the frame at the moment of discharge. This innovative system allowed for higher velocities and improved accuracy compared to other contemporary pistols. The toggle-lock design became synonymous with quality craftsmanship and precision engineering.

Another notable contribution by Georg Luger was his development of bottleneck cartridges for handguns. Prior to his innovation, most handguns used straight-walled cartridges that limited their ballistic performance. Luger's introduction of bottleneck cartridges significantly increased muzzle velocity while maintaining manageable recoil—a game-changer for firearm enthusiasts around the world.

Luger's dedication to perfection is evident in every aspect of his designs. He paid meticulous attention to details such as grip angle, trigger pull weight, and overall balance—factors that greatly influenced shooting comfort and accuracy. His commitment to user experience set new standards within the industry that are still upheld today.

Beyond his technical achievements, Georg Luger also played a pivotal role in shaping global firearm trends during World War I and World War II. The German military widely adopted his pistols during these conflicts due to their reliability, stopping power, and ease of maintenance—a testament to their exceptional performance on the battlefield. The Luger pistol became an iconic symbol of German engineering prowess and military might.

Although Georg Luger passed away in 1923, his contributions to firearms technology continue to influence modern designs. Many contemporary handguns owe their design principles and features to Luger's innovative ideas. His legacy lives on through firearms enthusiasts who appreciate the elegance and functionality of his creations.

Georg Luger's enduring legacy is built upon his revolutionary designs, including the iconic Luger pistol with its toggle-lock mechanism and bottleneck cartridges. His commitment to excellence, attention to detail, and contributions during times of conflict have solidified his position as a mastermind in firearms technology. The impact of his work can still be felt today, as many modern handguns draw inspiration from his groundbreaking innovations.

Georg Luger's name will forever be associated with quality craftsmanship and precision engineering in the world of firearms.

John Moses Browning

1855-1926

In the world of firearms, few names carry the weight and reverence that John Browning's does. Revered as a genius and celebrated as one of the most influential inventors in history, Browning's legacy has left an indelible mark on the field of firearms manufacturing. From his humble beginnings in Ogden, Utah, to his groundbreaking creations that revolutionized small arms technology, this text aims to delve into the life, works, and enduring impact of John Browning.

Born on January 23, 1855, in Ogden, Utah Territory, John Moses Browning was raised in a family with a deep-rooted passion for gunsmithing. His father, Jonathan Browning, was an accomplished gunsmith himself who had established a reputation for crafting exceptional rifles. Growing up amidst such familial expertise undoubtedly shaped young John's fascination with guns and ignited his own desire to push the boundaries of firearms design.

As a young man with an innate mechanical aptitude and relentless curiosity, Browning soon began tinkering with firearms designs in his father's workshop. It wasn't long before he started making significant breakthroughs that would lay the foundation for his future success. One early example was his development of a single-shot rifle mechanism at just age 23 – a concept that later evolved into one of his most iconic inventions: the Winchester Model 1885 Single Shot Rifle.

However impressive this achievement may have been at such a young age; it was only a glimpse into what would become an illustrious career characterized by prolific innovation. Throughout his life, John Browning patented over 128 different firearm designs – ranging from pistols to machine guns – making him one of history's most prolific inventors in this field. Beyond sheer quantity alone, though, lies the true genius and lasting impact of John Browning's inventions.

His designs were not just functional; they were innovative, reliable, and often ahead of their time. Whether it was the iconic Colt M1911 semi-automatic pistol, the Browning Automatic Rifle (BAR), or the revolutionary Browning Hi-Power pistol, each invention pushed the boundaries of firearms technology while simultaneously setting new standards for performance and reliability. Moreover, John Browning's contributions extended beyond individual inventions.

His innovative approach to firearm design fundamentally changed manufacturing processes and materials used in the industry. Many of his designs incorporated interchangeable parts, simplifying production and increasing efficiency – a concept that would later become standard practice in firearms manufacturing. As we delve deeper into the life and works of John Browning, it becomes clear that his legacy reaches far beyond his own achievements.

His inventions have shaped military strategies, influenced law enforcement practices, and even impacted popular culture through their depiction in movies and literature. Today, over a century after his passing in 1926 at age 71, John Browning's name remains synonymous with excellence in firearms design. In this text, we will explore the remarkable life story of John Moses Browning – from his early years as a budding inventor to his enduring influence on modern firearms technology.

***

John Browning, a name synonymous with innovation and excellence in the world of firearms, was born on January 23, 1855, in Ogden, Utah. From a young age, Browning displayed an extraordinary aptitude for mechanics and an insatiable curiosity about firearms. These early passions laid the foundation for his remarkable journey towards becoming one of the most influential gun inventors in history.

Growing up in a family deeply rooted in gunsmithing traditions, Browning was exposed to firearms from an early age. His father, Jonathan Browning, was an accomplished gunsmith himself and had established his reputation as a skilled craftsman. It was under his tutelage that John Browning began to develop both his technical skills and his love for all things related to firearms.

As a child, Browning spent countless hours assisting his father in their small workshop. He eagerly absorbed every aspect of gun design and manufacturing techniques. He studied the intricacies of different firearm mechanisms with great enthusiasm and demonstrated exceptional talent for understanding how these complex machines worked.

Browning's passion for firearms extended beyond mere mechanics; he possessed an innate ability to identify shortcomings in existing designs and tirelessly sought ways to improve upon them. This drive for innovation became apparent at a young age when he started sketching out his own firearm ideas on scraps of paper. These sketches were not mere fantasies but precursors to the revolutionary inventions that would later define his career.

In addition to being mechanically inclined, Browning also possessed an uncanny intuition when it came to understanding user needs and preferences. He recognized that guns should not only be reliable but also practical and comfortable to use. This insight set him apart from other inventors of his time who often focused solely on technical advancements without considering usability.

Browning's true genius began to emerge during his teenage years when he started developing fully functional prototypes based on his sketches. One of his earliest significant inventions was the single-shot rifle, which he designed at the age of 13. This remarkable achievement caught the attention of prominent gun manufacturers and collectors, who were astonished by the young prodigy's talent.

As Browning matured, his reputation as a firearms innovator continued to grow. He went on to invent numerous groundbreaking designs, including semi-automatic pistols, machine guns, and pump-action shotguns. His creations not only transformed the industry but also had a profound impact on military tactics and personal defense strategies.

The early life of John Browning was marked by an unwavering passion for firearms and an exceptional talent for mechanics. His upbringing in a family deeply involved in gunsmithing provided him with both the technical skills and the inspiration needed to become a true genius in his field. Through his relentless pursuit of innovation and dedication to user-centered design, Browning forever changed the landscape of firearms manufacturing and left an indelible mark on history.

***

John Moses Browning, an American firearms designer and inventor, is widely regarded as one of the most influential figures in the history of firearm design. His innovative breakthroughs revolutionized the field and set new standards for functionality, reliability, and efficiency. Browning's contributions spanned a wide range of firearms, from pistols to rifles and shotguns, leaving an indelible mark on the industry.

One of Browning's most significant breakthroughs came with his development of the recoil-operated firearm system. Prior to his invention, most firearms relied on manually operated mechanisms or gas-operated systems. However, Browning's innovation allowed for a more efficient and reliable mechanism that utilized the energy produced by firing a round to cycle the action. This breakthrough not only improved accuracy but also reduced recoil significantly, making firearms more comfortable and controllable for shooters.

Browning's genius extended beyond just inventing new systems; he was also known for his meticulous attention to detail in designing individual components. One notable example is his work on the 1911 pistol, which became one of his most iconic creations. The 1911 featured several groundbreaking design elements, such as a short recoil system with a tilting barrel that ensured reliable feeding and extraction while minimizing muzzle rise during firing.

Its single-action trigger mechanism sets new standards for crispness and consistency that still influence modern pistol designs today.

Another area where Browning made remarkable contributions was in shotgun design. He introduced numerous innovations that transformed shotguns from mere hunting tools into versatile firearms suitable for various applications. One such breakthrough was his creation of the first successful autoloading shotgun – the Auto-5 – which utilized long-recoil operation to cycle shells automatically while maintaining excellent reliability.

Moreover, Browning's designs were not limited to handguns and shotguns; he also made significant advancements in rifle technology. His lever-action rifles, such as the Winchester Model 1894, became immensely popular and set new standards for reliability and ease of use. Browning's lever-action designs were not only aesthetically pleasing but also featured ingenious mechanisms that allowed for quick reloading and smooth cycling of cartridges.

Apart from his technical innovations, Browning was known for his ability to create firearms that were both functional and aesthetically pleasing. His designs often incorporated elegant lines, graceful curves, and beautiful finishes that appealed to shooters and collectors alike. This attention to detail in both form and function earned him a reputation as a master craftsman.

John Browning's contributions to firearm design were nothing short of revolutionary. His innovative breakthroughs in recoil-operated systems, component design, shotgun technology, lever-action rifles, and overall aesthetics forever changed the landscape of firearms manufacturing. His legacy lives on through the countless models inspired by his work that are still cherished by shooting enthusiasts worldwide. John Moses Browning will always be remembered as a true pioneer whose ingenuity reshaped the world of firearms.

***

John Browning, often referred to as the "Father of Modern Firearms," was a prolific gun designer whose innovative creations revolutionized the industry. Throughout his career, Browning developed numerous iconic guns that not only changed the way firearms were designed but also had a lasting impact on military and civilian use. From handguns to machine guns, each of Browning's designs showcased his remarkable engineering skills and ingenuity.

One of Browning's most influential designs was the Colt M1911 pistol. Introduced in 1911, this semi-automatic handgun quickly became synonymous with reliability and power. The M1911 featured a unique short recoil system, which allowed for rapid firing while maintaining accuracy. Its design incorporated several groundbreaking features, including a single-action trigger mechanism and a detachable magazine. The Colt M1911 set the standard for future pistols and remains popular among firearm enthusiasts to this day.

Another groundbreaking invention by Browning was the Browning Automatic Rifle (BAR). Developed during World War I, this light machine gun was designed to provide infantrymen with portable firepower on the battlefield. The BAR featured an innovative gas-operated system that allowed for automatic fire while maintaining control and accuracy. It played a significant role in both World Wars and continued to be used by various military forces well into the 20th century.

Browning's contributions extended beyond handguns and machine guns; he also revolutionized the field of sporting rifles. One notable example is the Winchester Model 1894 lever-action rifle. Released in 1894, it became one of Winchester's most successful rifles due to its reliability, ease of use, and versatility in different shooting scenarios. Its lever-action mechanism allowed for quick follow-up shots, making it highly popular among hunters and sport shooters alike.

In addition to these remarkable designs, Browning also created iconic shotguns that are still revered today. The Auto-5, introduced in 1902, was the first successful semi-automatic shotgun. Its unique long-recoil system enabled rapid cycling and reduced felt recoil, setting a new standard for shotgun performance. The Auto-5 remained in production for almost 100 years and gained a reputation as one of the most reliable and versatile shotguns ever made.

Browning's genius lay not only in his innovative designs but also in his ability to adapt them to various applications. His firearms were not only admired by military forces but also embraced by civilians worldwide. Many of his designs continue to be manufactured and cherished by collectors, enthusiasts, and professionals alike.

John Browning's contributions to the field of firearms design are unparalleled. His iconic guns revolutionized the industry, setting new standards for reliability, functionality, and performance. From handguns like the Colt M1911 to machine guns such as the BAR, Browning's designs continue to shape modern firearms today. His legacy as a visionary engineer lives on through these influential creations that have left an indelible mark on the history of firearms design.

***

The name John Browning is synonymous with innovation and excellence in the world of firearms. His contributions to the industry have revolutionized the way firearms are designed and manufactured. One of his most significant partnerships was with the Winchester Repeating Arms Company, a collaboration that led to numerous groundbreaking inventions and forever changed the landscape of firearms. In 1883, John Browning entered into a partnership with Oliver Winchester, the owner of Winchester Repeating Arms Company.

This collaboration marked a pivotal moment in Browning's career and set the stage for his future successes. Together, they embarked on a journey to create some of the most iconic firearms in history. One of their earliest achievements was the development of the lever-action rifle. Lever-action rifles were already popular at that time, but Browning and Winchester sought to improve upon existing designs.

They introduced several key innovations that resulted in superior performance and reliability. The Model 1886 lever-action rifle was born out of this collaboration—a masterpiece that combined Browning's ingenious mechanisms with Winchester's manufacturing capabilities. Browning's inventive genius didn't stop there. In subsequent years, he continued to work closely with Winchester on various projects, each more ambitious than the last. One such milestone achievement was the creation of pump-action shotguns.

These shotguns became renowned for their smooth cycling action and exceptional reliability, thanks to Browning's innovative designs. Perhaps one of their most celebrated collaborations came in 1894 when they introduced one of America's most beloved rifles—the Model 1894 lever-action rifle chambered in .30-30 caliber. This rifle quickly gained popularity among hunters and sportsmen due to its compact size, lightweight construction, and remarkable accuracy.

As their partnership grew stronger over time, so did their commitment to pushing boundaries further. In 1910, they unveiled another groundbreaking invention—the first semi-automatic shotgun: The Auto-5. This design showcased Browning's ingenuity and Winchester's commitment to manufacturing excellence. The Auto-5 shotgun featured a unique long-recoil action, which delivered unmatched reliability and became the gold standard for semi-automatic shotguns. Throughout their collaboration, Browning and Winchester shared a common vision—to create firearms that were not only functional but also aesthetically pleasing.

Their designs incorporated elegant lines, intricate engravings, and high-quality craftsmanship that set them apart from their competitors. The partnership between John Browning and Winchester Repeating Arms Company left an indelible mark on the firearms industry. Their collaboration resulted in a long list of iconic firearms that continue to be cherished by collectors, enthusiasts, and sportsmen alike. Their legacy lives on through the enduring popularity of lever-action rifles, pump-action shotguns, and semi-automatic firearms—a testament to the enduring success of their milestone partnership.

John Browning's collaboration with Winchester Repeating Arms Company was a milestone partnership that revolutionized the world of firearms. Together, they created groundbreaking designs such as lever-action rifles, pump-action shotguns, and semi-automatic firearms that remain iconic to this day.

***

John Browning, a legendary American firearms designer, made an indelible mark on military history with his revolutionary machine guns. His innovative designs transformed the way wars were fought and greatly influenced the strategies and tactics employed by armed forces around the world. Browning's machine guns not only provided superior firepower but also increased mobility and versatility, forever changing the face of warfare.

One of Browning's most notable contributions to military history was the development of the M1917 water-cooled machine gun. Adopted by the United States Army during World War I, this weapon proved to be highly effective in combat situations. With its ability to sustain high rates of fire for extended periods without overheating, it gave American troops a significant advantage on the battlefield.

The M1917 became a staple of infantry units and played a crucial role in shaping offensive and defensive strategies during this conflict.

Browning's impact extended beyond his own country as well. His inventions were widely adopted by various nations around the world, enabling them to enhance their military capabilities. For instance, during World War II, both Allied and Axis powers utilized Browning-designed machine guns extensively. The reliability and firepower of these weapons allowed soldiers to suppress enemy positions effectively, providing cover for advancing troops or defending critical locations.

Perhaps one of Browning's most influential creations was the M2 .50 caliber machine gun or "Ma Deuce." This heavy machine gun became an iconic symbol of American military might throughout much of the 20th century and remains in service today. Its exceptional range, accuracy, and penetrating power made it an indispensable asset in both ground-based infantry operations and aerial combat scenarios.

Browning's inventions not only impacted battlefield tactics but also influenced broader military strategy. The introduction of reliable automatic weapons meant that armies had to adapt their approach to defensive positions such as trenches or fortified structures significantly. Machine guns could effectively suppress enemy infantry, making frontal assaults increasingly difficult and highlighting the importance of combined arms tactics to overcome these formidable defenses.

Furthermore, Browning's designs had a profound impact on the development of armored vehicles. The increased firepower provided by his machine guns necessitated the creation of better-protected vehicles that could withstand their devastating effects. Tanks and other armored fighting vehicles were equipped with Browning machine guns, enabling them to provide both anti-infantry support and anti-aircraft defense.

John Browning's revolutionary machine guns had a transformative effect on military history. His inventions provided unparalleled firepower, increased mobility, and versatility to armed forces around the world. From World War I through World War II and beyond, Browning's designs shaped battlefield tactics and influenced military strategies across various conflicts. His legacy lives on in modern warfare as his innovations continue to be utilized by armed forces globally.

***

John Moses Browning, often hailed as one of the most influential firearms designers in history, left an indelible mark on the industry with his innovative designs and numerous patents. Throughout his prolific career, Browning secured an impressive array of intellectual property rights, revolutionizing the world of firearms and forever changing the way we think about weapon design. Browning's inventive genius led to a staggering number of firearm patents that covered a wide range of innovative mechanisms and designs.

His first patent was issued in 1879 for a single-shot rifle design, marking the beginning of an illustrious journey into firearm innovation. Over the next four decades, he would amass an astonishing 128 firearm-related patents, each contributing to his legacy as a pioneer in weapons technology. One of Browning's most renowned inventions was the iconic Colt M1911 pistol. With its semi-automatic action and reliable .45 ACP caliber, it became synonymous with American military might during both World Wars.

Browning's patent for this pistol was issued in 1911 and is widely considered one of his most influential contributions to firearms history. Another notable patent from Browning is for the self-loading shotgun design known as the Auto-5 or Automatic Five. Introduced in 1902 by FN Herstal, this innovative shotgun featured a long-recoil operation that allowed for rapid firing while minimizing recoil impact on the shooter.

The Auto-5 became immensely popular and remained in production for almost a century after its introduction. Browning also held patents for various machine guns that played significant roles during major conflicts. One such example is his design for the M2 Browning machine gun, which has been used by militaries worldwide since its adoption by the U.S. Army in 1933. Known for its reliability and firepower, this heavy machine gun has become an integral part of modern warfare, mounted on vehicles, aircraft, and naval vessels.

Beyond handguns and machine guns, Browning's patents encompassed a wide range of firearm types. His designs included lever-action rifles like the Winchester Model 1886, pump-action shotguns such as the Remington Model 17, and even sporting rifles like the Browning BAR (Browning Automatic Rifle). Each of these patents contributed to advancements in their respective fields and solidified Browning's reputation as an unrivaled firearms innovator.

Browning's extensive intellectual property not only revolutionized firearms but also laid the foundation for subsequent generations of weapon designers. His patents continue to inspire and influence modern firearm development, with many of his designs still in use today. The enduring legacy of John Browning's intellectual property serves as a testament to his visionary approach and unwavering commitment to pushing the boundaries of firearm technology.

John Moses Browning's vast collection of firearm patents stands as a testament to his unparalleled contributions to the industry. From iconic pistols like the Colt M1911 to revolutionary shotguns such as the Auto-5, his innovative designs forever changed the landscape of firearms technology.

John Browning, a name synonymous with innovation and excellence in gun design, left an indelible mark on the firearms industry. His groundbreaking contributions revolutionized the way firearms were manufactured, enhancing their reliability, accuracy, and functionality. Browning's exceptional talent and unwavering commitment to his craft have earned him numerous honors and accolades throughout his career.

One of the most remarkable recognitions bestowed upon John Browning was his appointment as a technical expert by several governments around the world. His expertise in firearms design was sought after by nations seeking to modernize their military arsenals. From Belgium to France to Russia, Browning's skills were highly valued, leading him to collaborate with various governments on numerous projects.

In 1897, John Browning received the prestigious Chevalier of the Legion of Honor from France for his outstanding contributions to firearms design. This esteemed honor highlighted not only his exceptional talent but also recognized his role in shaping military strategies through weapon advancement.

Browning's ingenuity did not go unnoticed in his home country either. In 1926, he was awarded the United States Army Distinguished Service Medal for enhancing military capabilities through firearm innovations. This recognition demonstrated how deeply Browning's inventions had impacted national defense strategies.

In addition to government recognition, John Browning received accolades from renowned institutions within the firearm industry itself. The National Rifle Association (NRA) honored him posthumously in 2000 with their highest distinction: induction into their Hall of Fame. This recognition underlined how influential Browning had been in shaping American gun culture and further cemented his status as a legendary figure.

The arms manufacturing industry also paid tribute to John Browning's brilliance by naming multiple awards after him. The "John M. Browning Outstanding Achievement Award" is presented annually by both the Sporting Arms and Ammunition Manufacturers' Institute (SAAMI) and the National Association of Sporting Goods Wholesalers (NASGW). These awards serve as a testament to Browning's enduring impact on the design and production of firearms.

John Browning's legacy is not confined to awards and honors alone. His designs have become timeless classics, beloved by shooters, collectors, and enthusiasts worldwide. The Winchester Model 1894 lever-action rifle, the Colt M1911 semi-automatic pistol, and the Browning Auto-5 shotgun are just a few examples of his iconic creations that remain highly sought-after to this day.

Furthermore, many firearm manufacturers continue to produce firearms based on John Browning's designs. Companies like FN Herstal, Browning Arms Company, and Winchester Repeating Arms have dedicated themselves to preserving his legacy by manufacturing firearms that embody his innovative spirit.

John Browning's contributions to gun design were so groundbreaking and influential that they earned him numerous honors throughout his career. From being appointed as a technical expert by governments worldwide to receiving prestigious accolades from institutions within the industry itself, Browning's exceptional talent was recognized far and wide. Beyond the honors he received during his lifetime, his enduring legacy lives on through iconic firearms still cherished by enthusiasts today.

**

John Browning is widely regarded as one of the most influential figures in the history of firearms manufacturing and design. His innovative ideas and engineering prowess revolutionized the industry, leaving an indelible mark on modern firearms manufacturing techniques and design principles.

One of Browning's most significant contributions was his pioneering work on automatic firearms. He was a trailblazer in developing reliable, self-loading mechanisms that could feed ammunition into the chamber without manual intervention. His designs, such as the famous Browning Automatic Rifle (BAR) and the Colt Model 1911 pistol, set new standards for efficiency and reliability in automatic weapons.

Browning's approach to firearm manufacturing focused on simplicity, durability, and ease of use. He believed that a firearm should be easy to operate even under harsh conditions. To achieve this goal, he incorporated ingenious mechanisms into his designs that reduced complexity while enhancing reliability. For example, his tilting barrel mechanism used in many pistols allowed for more efficient recoil management while simplifying construction.

Moreover, Browning had a profound impact on firearm ergonomics. He understood the importance of creating weapons that were comfortable to handle and could be effectively operated by shooters of different sizes and physical abilities. His designs featured ergonomic grips, optimized trigger pulls, and intuitive control layouts. These principles continue to influence modern firearm design by prioritizing user comfort and effective handling.

Browning also introduced several groundbreaking manufacturing techniques that are still widely used today. One such technique was his innovative use of stamped metal parts instead of traditional forged components for certain firearm assemblies. This not only reduced production costs but also increased efficiency without compromising durability or performance.

Furthermore, Browning played a pivotal role in advancing cartridge technology through his development of various calibers for both handguns and rifles. His designs were optimized for maximum power transfer from the cartridge to propel projectiles with exceptional velocity while maintaining manageable recoil levels. This emphasis on ballistic performance continues to shape modern ammunition design principles.

In addition to his technical contributions, Browning's business acumen and partnerships with prominent firearm manufacturers greatly influenced the industry's landscape. His collaborations with Winchester Repeating Arms Company and Fabrique Nationale (FN) resulted in the successful production of many of his designs, further cementing his legacy as a master gunsmith.

John Browning's influence on modern firearms manufacturing techniques and design principles cannot be overstated. His innovative ideas, commitment to simplicity and reliability, focus on ergonomics, introduction of new manufacturing techniques, advancements in cartridge technology, and successful partnerships all contributed to his lasting impact on the industry. Today, firearms manufacturers continue to draw inspiration from Browning's work as they strive to create ever-improving weapons that uphold his legacy of excellence.

***

John Moses Browning, a name synonymous with innovation and excellence in the world of firearms, left an indelible mark on history as the mastermind behind countless legendary guns. His relentless pursuit of perfection, unmatched engineering skills, and unwavering passion for his craft revolutionized the firearms industry and shaped the way we perceive modern weaponry. As we reflect on his remarkable contributions, it becomes evident that John Browning's legacy will forever be remembered as an icon of ingenuity.

Throughout his illustrious career spanning several decades, Browning introduced a plethora of groundbreaking designs that continue to influence firearm development today. From iconic handguns like the Colt M1911 to iconic long arms such as the Winchester Model 1894 and Browning Automatic Rifle (BAR), his creations have become symbols of power, reliability, and precision. These weapons not only transformed warfare but also became beloved tools for sport shooting enthusiasts worldwide.

One cannot discuss John Browning's genius without acknowledging his unparalleled attention to detail. Known for meticulously refining each design until it reached perfection, he prioritized functionality over everything else. With an astute understanding of mechanics combined with artistic craftsmanship, he created firearms that not only performed flawlessly but also possessed an aesthetic appeal unrivaled by their contemporaries.

Browning's relentless pursuit of innovation pushed boundaries in both form and function. He was never content with resting on past achievements; instead, he constantly sought new ways to improve upon existing designs or create entirely new ones altogether. From pioneering gas-operated systems to inventing recoil-operated mechanisms and introducing detachable box magazines – each invention demonstrated his unrivaled ability to anticipate future needs while staying ahead of technological advancements.

Beyond his technical brilliance, John Browning's impact on society extended beyond engineering feats. His designs played a significant role in shaping world history by equipping military forces during critical moments and empowering individuals with reliable self-defense options. The sense of security his firearms provided was immeasurable, instilling confidence in those who wielded them.

Even today, Browning's influence persists in the firearm industry. His legacy is carried forward by the Browning Arms Company, which continues to produce high-quality firearms that embody his principles of excellence and innovation. The company's commitment to upholding the standards set by its founder ensures that future generations will continue to benefit from his ingenuity.

John Moses Browning's remarkable contributions as a mastermind behind countless legendary guns cannot be overstated. His relentless pursuit of perfection, unmatched engineering skills, and unwavering passion for his craft have left an indelible mark on history. From handguns to long arms, each design he created showcased not only technical brilliance but also aesthetic appeal. As we remember him today, it is clear that John Browning's genius forever changed the face of firearms manufacturing and will continue to inspire future generations of engineers and enthusiasts alike.

John Thompson
1860-1940

The evolution of machine guns has played a pivotal role in shaping the history of warfare, as these weapons have continuously pushed the boundaries of technological innovation. Tracing their origins reveals a remarkable journey from primitive hand-cranked mechanisms to the high-powered automatic firearms we see on today's battlefields.

The concept of a machine gun can be traced back to ancient times, where early civilizations experimented with various forms of rapid-fire weapons. However, it was not until the late 19th century that significant advancements were made. The first true machine gun was patented by Richard Gatling in 1862, and it featured multiple barrels revolving around a central axis, enabling sustained fire.

Following Gatling's invention, several inventors sought to improve upon his design. One such innovator was Hiram Maxim, who introduced the world to the recoil-operated machine gun in

1883. This mechanism harnessed the energy generated by each round fired to automatically reload and cock the weapon for subsequent shots. Maxim's invention marked a major leap forward in firepower and set the stage for further developments.

In World War I, machine guns became integral components of military strategies due to their ability to deliver devastating firepower. The introduction of water-cooled systems allowed for prolonged firing without overheating, while air-cooled variants offered increased mobility on the battlefield. Despite these advancements, however, early machine guns still relied on manual operation and required considerable manpower.

It wasn't until World War II that significant breakthroughs occurred in automation technology. John Thompson's legendary submachine gun played a crucial role during this period by incorporating innovative features that revolutionized combat tactics. Thompson's weapon utilized an open-bolt design coupled with blowback operation—a mechanism that used propellant gases generated by firing each round to push back against the bolt and chamber another cartridge automatically.

Furthermore, Thompson's submachine gun featured select-fire capability—allowing users to switch between semi-automatic and fully automatic modes. This versatility gave soldiers the ability to adapt their firepower to different combat scenarios, making it an incredibly effective weapon in close-quarters engagements.

Advancements in materials and manufacturing techniques during the post-World War II era further propelled the evolution of machine guns. The introduction of lightweight alloys and improved machining processes led to the development of more compact and portable firearms without compromising firepower. These advancements paved the way for modern-day machine guns such as the M249 Squad Automatic Weapon (SAW) and the FN Minimi, which are widely used by military forces around the world.

Tracing the origins of machine gun innovation reveals a fascinating journey characterized by leaps in technology and a constant quest for increased firepower and efficiency. From Gatling's hand-cranked mechanism to Thompson's submachine gun, these weapons have evolved significantly over time, shaping warfare as we know it today. The relentless pursuit of innovation on the battlefield continues to drive advancements in machine gun technology, ensuring that future generations will witness even more remarkable developments in this field.

***

In the annals of military history, few weapons have achieved the legendary status and lasting impact of the Thompson submachine gun. Designed by John T. Thompson, a brilliant engineer and entrepreneur, this remarkable firearm revolutionized warfare with its unprecedented firepower and versatility. The story behind the creation of the Thompson submachine gun is one of innovation, determination, and a deep understanding of the needs of soldiers on the battlefield.

At a time when World War I was wreaking havoc across Europe, John Thompson recognized the limitations of traditional firearms in close-quarter combat situations. He saw an opportunity to develop a weapon that could provide superior firepower while being compact enough for mobility. Thus began his quest to create what would become known as "The Tommy Gun."

Thompson's design philosophy centered around creating a lightweight and portable firearm that could deliver rapid-fire capabilities. He envisioned a weapon that would bridge the gap between pistols and rifles, offering soldiers an unprecedented advantage in close-range encounters. To achieve this vision, he focused on developing an automatic firearm that utilized pistol ammunition.

The result was nothing short of revolutionary. The Thompson submachine gun featured a distinctive appearance with its iconic drum magazine mounted on top—an innovation that allowed for continuous firing without reloading frequently. With its ability to fire up to 1,500 rounds per minute, this weapon provided unparalleled firepower to infantry units.

One key element that set the Thompson submachine gun apart from other firearms was its select-fire capability. Soldiers could choose between fully automatic fire or single-shot mode depending on their tactical needs—a feature unheard of at that time. This adaptability allowed troops to switch seamlessly between suppressive fire during assaults or precise shots when engaging individual targets.

Moreover, Thompson's attention to detail extended beyond just designing a powerful weapon; he also emphasized user comfort and ease-of-use. The Tommy Gun boasted an ergonomic design featuring a forward pistol grip, a shoulder stock for stability, and a low recoil mechanism. These features made it easier for soldiers to handle the weapon effectively, even during extended periods of combat.

The impact of the Thompson submachine gun on warfare cannot be overstated. Its introduction forever changed the dynamics of close-quarter combat, giving soldiers an unprecedented advantage in firepower and versatility. The Tommy Gun became particularly renowned during World War II when it found its way into the hands of soldiers across different theaters of war.

The Thompson submachine gun was a true game-changer in the realm of firepower revolution. John Thompson's unwavering commitment to innovation and his understanding of the needs of soldiers led to the creation of a weapon that remains iconic to this day. The remarkable story behind this legendary firearm serves as a testament to human ingenuity and its ability to shape history on the battlefield.

***

In the annals of military history, certain individuals stand out for their innovative contributions that forever changed the face of warfare. Among these visionaries is John Thompson, an inventor extraordinaire whose remarkable story is intertwined with the creation of his legendary submachine gun. Thompson's journey from humble beginnings to becoming a renowned inventor and his relentless pursuit of innovation on the battlefield are nothing short of extraordinary.

Born in 1860 in Newport, Kentucky, John Thompson exhibited a natural curiosity and passion for mechanics from an early age. His insatiable appetite for knowledge led him to pursue engineering at Purdue University, where he honed his skills and developed a keen understanding of machinery. After graduating with top honors, Thompson embarked on a career that would shape the course of military technology forever.

Thompson's first breakthrough came when he joined the United States Army Ordnance Department in 1886. Recognizing his exceptional talent and determination, he quickly rose through the ranks and became known for his ability to solve complex engineering problems. However, it was during World War I that Thompson truly shone as an inventor.

Witnessing the devastating stalemate caused by trench warfare on the Western Front, Thompson became determined to develop a weapon that could provide soldiers with increased firepower while maintaining mobility. This led him to conceptualize what would later become known as the "Thompson Submachine Gun" or colloquially referred to as "the Tommy Gun."

Thompson's submachine gun was revolutionary in several aspects. Its compact size allowed soldiers greater maneuverability in close-quarter combat while delivering rapid-fire capabilities with deadly accuracy. The weapon's recoil system further enhanced its stability, ensuring accurate targeting even during sustained bursts.

However, turning his visionary design into reality was no easy task. Overcoming numerous technical challenges and financial constraints, Thompson founded Auto-Ordnance Corporation in 1916 to develop and manufacture his groundbreaking invention. Despite setbacks and a lengthy development process, Thompson's unwavering determination eventually paid off when the first production models of the submachine gun were introduced in 1921.

The impact of Thompson's submachine gun on military tactics was profound. It soon gained popularity among law enforcement agencies and armed forces worldwide, becoming an iconic symbol of power and innovation. Its effectiveness in close-quarter combat saw it deployed across various theaters during World War II, where it played a pivotal role in shaping the outcome of numerous battles.

John Thompson's legacy as an inventor extraordinaire extends far beyond his remarkable submachine gun. His relentless pursuit of innovation on the battlefield not only revolutionized military technology but also saved countless lives by empowering soldiers with superior firepower. Thompson's extraordinary story serves as a testament to the power of human ingenuity and its ability to shape history.

***

Innovation has always been a driving force behind military advancements throughout history. The desire for more effective weapons and strategies has led to the development of remarkable technologies that have reshaped the battlefield. One such innovation that stands out is the legendary submachine gun created by John Thompson, a visionary inventor whose determination and ingenuity brought his concept to reality.

John Thompson's journey began in the early 20th century when he recognized a pressing need for a compact firearm that could provide rapid-fire capability while remaining portable. Inspired by the challenges faced by soldiers during World War I, Thompson envisioned a weapon that would revolutionize close-quarters combat. With this vision in mind, he embarked on a quest to design and build what would become one of the most iconic firearms in history.

Thompson's first step was to assemble a team of skilled engineers who shared his passion for innovation. Together, they set out to create an entirely new type of firearm – one that combined portability with firepower. Countless hours were spent brainstorming ideas, refining designs, and conducting rigorous testing. It was crucial for Thompson's team not only to meet but exceed expectations in terms of reliability, accuracy, and ease of use.

The process from concept to reality was no easy feat. The team encountered numerous challenges along the way – from technical hurdles to financial constraints. However, their unwavering dedication and belief in their vision kept them going during times of adversity.

Thompson understood that success required not only technical excellence but also effective marketing strategies. He tirelessly promoted his invention through demonstrations and presentations at military bases across the country. His efforts caught the attention of key decision-makers who recognized the potential value his submachine gun could bring to modern warfare.

Once Thompson secured essential funding from investors who shared his vision, production began at full speed.

Manufacturing facilities were established, and an assembly line was set up – marking yet another milestone in the journey from concept to reality. Skilled workers meticulously crafted each component, ensuring the highest level of quality.

As the first Thompson submachine guns rolled off the assembly line, they were met with great anticipation by soldiers and military officials alike. The firearm's exceptional performance in both close quarters combat and suppression fire scenarios surpassed expectations. Its innovative design, featuring a high-capacity magazine and selective fire capability, made it a force to be reckoned with on the battlefield.

The impact of Thompson's innovation cannot be overstated. It forever changed the way wars were fought, giving soldiers a distinct advantage in close-range encounters. Its influence was felt not only during World War II but also in subsequent conflicts worldwide.

John Thompson's remarkable story serves as an inspiration for future generations of innovators who strive to unleash their ideas onto the battlefield. His unwavering dedication, technical expertise, and relentless pursuit of perfection transformed an ambitious concept into a reality that shaped military history. The legendary submachine gun he created stands as a testament to the power of innovation and its profound impact on warfare.

***

In the annals of military history, few weapons have achieved legendary status like the submachine gun. A weapon that revolutionized warfare, it provided soldiers with unprecedented firepower and mobility on the battlefield. Among the pioneers who contributed to this remarkable invention was John Thompson, whose story and his iconic submachine gun continue to captivate military enthusiasts worldwide.

John Thompson, an engineer and inventor, embarked on a mission to develop a weapon that would bridge the gap between pistols and rifles during World War I. With an innate understanding of firearms and a relentless pursuit of innovation, Thompson set out to create a compact yet potent firearm that would give soldiers a decisive advantage in close-quarter combat.

After years of tireless experimentation and refinement, Thompson unveiled his masterpiece in 1919—the first fully operational prototype of what would later become known as the Thompson submachine gun. Nicknamed "The Annihilator" by its creator due to its impressive firepower, this submachine gun was unlike anything seen before.

The legendary submachine gun combined several innovative features that set it apart from existing firearms. One defining characteristic was its automatic firing capability—allowing for rapid-fire bursts with minimal recoil—a feature previously unseen in small arms. This groundbreaking innovation made it ideal for trench warfare or any situation where speed and accuracy were paramount.

Thompson's design also boasted an extended magazine capacity compared to other contemporary firearms. The detachable drum magazine could hold up to fifty rounds, providing soldiers with sustained firepower without frequent reloading—a crucial advantage in intense combat scenarios.

Furthermore, Thompson paid meticulous attention to ergonomics when designing his legendary firearm. The well-balanced weight distribution and comfortable grip allowed for easy maneuverability even during prolonged engagements. The weapon's compactness made it suitable for use in tight spaces such as urban environments or armored vehicles—an asset that quickly endeared it to soldiers across different theaters of war.

The Thompson submachine gun's impact was immediate and far-reaching. It quickly gained popularity among law enforcement agencies, military units, and even notorious gangsters during the Prohibition era in the United States. Its reputation as a highly effective weapon solidified its status as an icon of power and authority.

Despite evolving firearms technology over the decades, the Thompson submachine gun's legacy endures. It remains a symbol of innovation, resilience, and ingenuity—qualities that define John Thompson himself. His relentless pursuit of perfection in firearm design revolutionized military tactics and established a new standard for small arms manufacturing.

The birth of the legendary submachine gun represents a pivotal moment in military history. John Thompson's remarkable invention changed the face of warfare by providing soldiers with an unprecedented combination of firepower, mobility, and reliability. The timeless appeal and enduring legacy of this iconic firearm continue to captivate enthusiasts around the world—an immortal testament to one man's quest for innovation on the battlefield.

***

The Thompson submachine gun, also known as the Tommy Gun, revolutionized warfare tactics when it was introduced on the battlefield. This iconic weapon played a pivotal role in transforming the way wars were fought, offering unprecedented firepower and versatility to soldiers. Under the guidance of John Thompson, this legendary submachine gun forever changed the course of military history.

One of the most significant contributions of the Thompson submachine gun was its ability to deliver a high volume of automatic fire. Unlike traditional rifles that required manual reloading after each shot, this weapon could fire continuous bursts with its large-capacity drum magazine or detachable box magazine. This rapid-fire capability allowed soldiers to suppress enemy positions and gain a significant advantage in combat situations.

Moreover, the Thompson's compact size and portability made it ideal for close-quarter battles. In urban warfare scenarios or tight spaces like trenches and buildings, where traditional rifles were impractical to maneuver, this submachine gun excelled. Its short barrel length enabled easy handling and swift target acquisition in confined environments.

The exceptional performance of the Thompson submachine gun also led to new tactical approaches on the battlefield. The weapon's effectiveness in providing cover fire allowed troops to advance under intense enemy fire while minimizing casualties. By suppressing adversaries with sustained automatic fire, soldiers could create windows of opportunity for their comrades to maneuver and gain strategic advantages.

Furthermore, its unique design influenced infantry tactics by introducing a new concept: squad-level firepower superiority. Small units equipped with Thompson submachine guns could unleash devastating firepower against larger enemy forces while maintaining mobility and flexibility on the battlefield. This shift from individual marksmanship towards collective firepower fundamentally transformed infantry tactics during World War II.

The versatility of this iconic firearm extended beyond conventional warfare as well. The Thompson found utility in unconventional operations such as sabotage missions or guerrilla warfare due to its small size and ease of concealment. Its use by resistance fighters against occupying forces during World War II further exemplifies how this weapon became a symbol of resistance and liberation.

The Thompson submachine gun had a profound impact on warfare tactics, forever changing the way battles were fought. Its high rate of fire, portability, and effectiveness in close-quarter combat provided soldiers with an unprecedented advantage on the battlefield. Under the guidance of John Thompson, this legendary firearm transformed infantry tactics by emphasizing squad-level firepower superiority. Whether used in conventional or unconventional operations, the Thompson submachine gun became an emblem of innovation and resilience during times of conflict.

***

The invention of the submachine gun by John Thompson revolutionized warfare and forever altered the dynamics on the battlefield. With its remarkable design and unprecedented firepower, this legendary weapon became a game-changer, shifting military strategies and tactics while leaving an indelible mark on history.

The submachine gun's impact on warfare was most evident during World War II. Prior to its introduction, soldiers primarily relied on bolt-action rifles, which had limited firing rates and were ill-suited for close-quarter combat. However, Thompson's submachine gun, commonly known as the "Tommy Gun," changed this paradigm entirely.

One of the key features that set Thompson's submachine gun apart was its automatic firing capability. Unlike previous firearms that required manual cocking and reloading after each shot, this weapon could fire continuously with a high rate of fire. This innovation gave soldiers an unprecedented advantage in close-range engagements, enabling them to unleash a torrent of bullets upon their enemies.

The Tommy Gun's firepower made it particularly effective in urban warfare scenarios. In dense city streets or confined spaces such as buildings or trenches, traditional rifles were cumbersome and slow to operate. The compact design of Thompson's submachine gun allowed for greater mobility and maneuverability in these environments. Soldiers armed with Tommy Guns could rapidly engage multiple targets at close range without compromising their own safety.

Moreover, the psychological impact of this weapon cannot be overstated. The distinct sound produced by its firing mechanism – often described as a "rat-a-tat" – struck fear into enemy forces who had never encountered such a relentless barrage of bullets before. This auditory intimidation factor further contributed to the weapon's dominance on the battlefield.

The submachine gun also played a pivotal role in shaping military tactics during World War II. Its effectiveness led to new strategies emphasizing rapid assaults and overwhelming firepower. Infantry units armed with Tommy Guns became specialized assault teams, capable of quickly breaching enemy defenses and neutralizing fortified positions. The submachine gun's ability to provide suppressive fire allowed for the coordination of complex maneuvers, leading to the eventual development of squad-based tactics still in use today.

Beyond its impact during World War II, the submachine gun's legacy continued to shape future conflicts and military doctrines. Its influence can be seen in subsequent designs such as the Soviet Union's PPSh-41 and Germany's MP40, both of which aimed to replicate the success of Thompson's invention. Additionally, advancements in firearms technology owe much to Thompson's pioneering work on automatic weapons.

John Thompson's legendary submachine gun revolutionized warfare by providing soldiers with unprecedented firepower and mobility. Its introduction fundamentally changed military tactics and strategies during World War II and beyond. The submachine gun remains an enduring symbol of innovation on the battlefield, forever altering the course of history.

***

John Thompson, a name synonymous with innovation and excellence in the field of weapon design, left an indelible mark on the battlefield with his remarkable submachine gun. His invention revolutionized warfare, providing soldiers with a powerful and reliable weapon that would change the course of history. Today, we honor Thompson's invaluable contribution to weapon innovation and acknowledge the lasting legacy he has left behind.

Thompson's journey towards creating his legendary submachine gun began during the tumultuous years of World War I. Witnessing the devastating impact of trench warfare firsthand, he recognized the need for a firearm that could provide both firepower and mobility. With unwavering determination, Thompson set out to design a weapon that would meet these demands.

After several years of research and development, Thompson unveiled his masterpiece - the Thompson submachine gun or "Tommy Gun." Featuring a compact size, high rate of fire, and exceptional reliability, this innovative weapon quickly gained recognition among military circles. Its effectiveness on the battlefield was unmatched at that time.

During World War II, thousands of American soldiers depended on the Thompson submachine gun as they fought against Axis forces in Europe and Asia. Its iconic appearance became synonymous with American military might. The Tommy Gun provided soldiers with unparalleled firepower in close quarters combat situations such as urban warfare or clearing enemy bunkers. It was highly regarded for its accuracy and ease of use even under adverse conditions.

Beyond its battlefield success, Thompson's submachine gun had a profound impact on subsequent generations of firearms designers. Its innovative blowback operation system influenced many future weapons' designs across various nations around the world. The legacy of this groundbreaking firearm lives on in countless modern firearms used by armed forces worldwide.

Moreover, John Thompson's contribution extended beyond just inventing an exceptional weapon; he also played a significant role in revolutionizing manufacturing processes for small arms production. He introduced mass production techniques to meet the high demand for his submachine gun during World War II. Thompson's innovative production methods greatly influenced subsequent firearm manufacturing, making it faster and more efficient.

Today, we honor John Thompson's remarkable achievements by recognizing him as a pioneering figure in weapon innovation. His contribution to the field of firearms design continues to shape modern warfare. The impact of his submachine gun on the battlefield cannot be overstated, as it provided soldiers with a powerful tool that significantly enhanced their capabilities.

As we reflect upon Thompson's legacy, we are reminded of the importance of innovation and its ability to shape the course of history. His dedication to improving soldiers' lives on the battlefield serves as an inspiration for future generations of weapon designers and military strategists.

John Thompson's remarkable story and his legendary submachine gun have left an enduring legacy in the realm of weapon innovation. By honoring his contributions, we pay tribute not only to a brilliant mind but also to all those who have dedicated their lives to advancing military technology for the greater good.

***

The Thompson submachine gun, also known as the Tommy Gun, is a legendary weapon that has left an indelible mark on military history. Its iconic design, rapid-fire capabilities, and reliability made it a favorite among soldiers and law enforcement officers alike. But what were the secrets behind its unparalleled success?

One of the key factors behind the triumph of the Thompson submachine gun was its innovative blowback-operated system. Designed by John Thompson himself, this system utilized energy from fired rounds to cycle the action and load subsequent rounds into the chamber. This not only ensured smooth firing but also allowed for a high rate of fire without sacrificing accuracy.

Another secret to its success was its adaptability. The Thompson submachine gun was designed to be versatile, allowing it to serve in various roles on the battlefield. It could be fired from either shoulder or hip, making it ideal for close quarters combat such as trench warfare or clearing buildings during urban operations. Its compact size and lightweight nature made it easy to handle and maneuver in tight spaces.

Furthermore, a unique feature that set the Thompson submachine gun apart was its drum magazine. Capable of holding up to 50 or 100 rounds depending on variants, this magazine provided an impressive firepower advantage over other contemporary firearms. The drum magazine's capacity allowed soldiers to sustain suppressive fire for longer periods without reloading frequently, giving them a significant tactical advantage.

Reliability played a crucial role in establishing the reputation of the Thompson submachine gun as well. Soldiers needed weapons they could trust in life-or-death situations, and John Thompson understood this perfectly. He meticulously engineered every aspect of his firearm to ensure durability under extreme conditions.

Additionally, one cannot overlook how ergonomics contributed to its success. The Thompson submachine gun featured an ergonomic grip and stock design that allowed for comfortable handling even during prolonged use. This attention to detail made it easier for soldiers to maintain control and accuracy while firing, enhancing their overall effectiveness on the battlefield.

Moreover, the Thompson submachine gun's success can also be attributed to its exceptional craftsmanship. Each weapon was meticulously manufactured with precision and attention to detail. The quality of materials used in its construction, coupled with expert craftsmanship, ensured that every Thompson submachine gun delivered superior performance.

The secrets behind the remarkable success of the Thompson submachine gun can be attributed to several key factors. Its innovative blowback-operated system, adaptability, high-capacity drum magazine, reliability under adverse conditions, ergonomic design, and exceptional craftsmanship all contributed to its legendary status. John Thompson's dedication to creating a weapon that surpassed expectations led to an innovation that revolutionized warfare and left an enduring legacy in military history.

***

John Thompson, a remarkable innovator and inventor, left an indelible mark on the world with his legendary submachine gun. His story not only showcases the power of determination and perseverance but also provides valuable lessons for aspiring future innovators. Thompson's revolutionary creation offers insights into the mindset and qualities required to unleash innovation on any battlefield.

Firstly, John Thompson's story teaches us that innovation often stems from a deep understanding of a problem or need. As a former soldier himself, Thompson recognized the limitations of existing firearms during World War I. He saw firsthand how soldiers struggled in close-quarter combat situations, which led him to envision a weapon that would bridge the gap between rifles and handguns.

This understanding allowed him to identify an opportunity for improvement and set out on his journey towards creating something extraordinary.

Furthermore, Thompson demonstrates that true innovation requires relentless curiosity and continuous learning. Throughout his development process, he experimented with various designs and technologies before settling on the iconic submachine gun we know today. He tirelessly sought feedback from soldiers in the field and collaborated with experts to refine his creation further. By embracing curiosity and constantly seeking knowledge, Thompson was able to push boundaries beyond what was thought possible at the time.

Another crucial lesson from John Thompson's innovation is the importance of collaboration and teamwork. While he played a central role in designing the submachine gun, he understood that success could only be achieved through collective effort. Thompson assembled a team of skilled engineers who shared his vision and worked tirelessly alongside him to bring it to life. Their combined expertise enabled them to overcome challenges that would have been insurmountable for an individual working alone.

Moreover, Thompson's journey reminds us that failure is an inherent part of any innovative process but should never deter one from pursuing their goals. Despite facing setbacks such as funding shortages or technical difficulties, he remained steadfast in his pursuit of excellence. Thompson's ability to learn from failures, adapt his approach, and persist in the face of adversity is a testament to his unwavering commitment to his mission.

Lastly, John Thompson's story serves as a reminder that innovation is not solely driven by personal ambition but can have a profound impact on society. His submachine gun revolutionized warfare tactics and saved countless lives on the battlefield. Thompson's determination to improve the lives of soldiers exemplifies the importance of aligning innovation with a higher purpose or societal need.

John Thompson's remarkable story provides aspiring future innovators with valuable lessons that transcend time. From understanding the problem at hand to embracing curiosity, collaboration, resilience, and purpose-driven innovation, Thompson's journey offers a roadmap for those seeking to unleash their own creativity and make a lasting impact on the world. By internalizing these lessons and embodying the spirit of innovation demonstrated by John Thompson, future innovators can pave their own path toward shaping a better tomorrow.

Hugo Schmeisser
1884-1953

Hugo Schmeisser, a renowned German engineer, is widely regarded as one of the most influential figures in the field of firearms design. His innovative contributions revolutionized weapon technology, particularly in the development of rifles and submachine guns. Born on September 24, 1884, in Jena, Germany, Schmeisser's life was dedicated to pushing the boundaries of firearm engineering and leaving an indelible mark on military weaponry.

From an early age, Schmeisser displayed a keen interest in mechanics and engineering. He was born into a family with a long-standing tradition in the arms industry; his father, Louis Schmeisser was a respected firearms designer himself. Growing up surrounded by tools and prototypes undoubtedly fueled young Hugo's passion for this craft.

In his early twenties, Hugo Schmeisser joined forces with his father at their family-owned factory in Suhl. This collaboration allowed him to acquire invaluable hands-on experience and mentorship from his father. Under Louis' guidance, Hugo honed his skills as a gunsmith while simultaneously developing an understanding of manufacturing processes.

Schmeisser's breakthrough came in 1915 when he unveiled his masterpiece: the MP18 submachine gun. This weapon would forever change military tactics by introducing fully automatic fire capabilities to individual soldiers. The MP18 became widely recognized as one of the first practical submachine guns ever produced and played a crucial role during World War I.

Following this success, Schmeisser continued to innovate throughout his career, leaving an indelible mark on firearm design. He consistently pushed boundaries by experimenting with unconventional materials such as polymers and integrating cutting-edge technologies into weapon systems.

Schmeisser's pioneering work extended beyond submachine guns; he also made significant contributions to rifle design. One notable example is the StG44 (Sturmgewehr 44), often referred to as the world's first assault rifle. This groundbreaking firearm combined the firepower of a machine gun with the portability and ease of use of a rifle, thus revolutionizing infantry tactics. The StG44 had a lasting impact on military strategies and influenced the design of subsequent assault rifles worldwide.

Unfortunately, Schmeisser's career took an unfortunate turn after World War II when he was captured by Soviet forces and taken to their research facilities in Izhevsk. Forced to work under duress, Schmeisser was involved in designing firearms for the Soviet Union during his captivity. However, despite these circumstances, his contributions to weapon technology cannot be denied.

Hugo Schmeisser's life and work as a German engineer were characterized by groundbreaking innovations in firearms design. From his early days at his family's factory to his influential creations like the MP18 submachine gun and StG44 assault rifle, Schmeisser left an indelible mark on military weaponry. Despite facing adversity during World War II, Schmeisser's legacy endures as a testament to his genius and unwavering dedication to pushing the boundaries of engineering excellence in firearms design.

***

Hugo Schmeisser, a German engineer, was a revolutionary figure in the development of automatic weapons during the early 20th century. From rifles to submachine guns, his contributions have left an indelible mark on the history of firearms. Schmeisser's innovative designs and engineering prowess transformed traditional firearms into highly efficient and deadly automatic weapons.

One of Schmeisser's notable contributions was his work on improving the functionality and reliability of rifles. He introduced several groundbreaking features that enhanced their performance on the battlefield. One such innovation was the introduction of detachable box magazines. Before this advancement, rifles were loaded with individual cartridges through a time-consuming process. Schmeisser's detachable magazine allowed soldiers to load multiple rounds quickly, significantly increasing their rate of fire during combat.

Schmeisser also played a crucial role in developing selective-fire mechanisms for rifles. This feature allowed soldiers to switch between semi-automatic and fully automatic firing modes with ease. By enabling rapid bursts of fire, these rifles gave infantry units a significant advantage in combat situations where suppressing enemy positions was critical.

However, it was with submachine guns that Hugo Schmeisser truly showcased his genius as an engineer. The submachine gun represented a paradigm shift in firearms design, providing soldiers with compact yet powerful automatic weapons suitable for close-quarter combat.

Schmeisser's most famous creation, the MP18 submachine gun, revolutionized warfare during World War I. This weapon boasts impressive firepower while maintaining maneuverability due to its lightweight design. The MP18 featured an open bolt system that contributed to its reliability by preventing overheating during sustained fire.

Another notable contribution by Schmeisser was his pioneering use of intermediate cartridges for submachine guns. These cartridges struck a balance between rifle rounds' power and handgun ammunition's size, making them ideal for close-quarters engagements while still maintaining lethality at longer ranges. This innovation laid the foundation for future developments in firearms design, including the creation of assault rifles.

Schmeisser's engineering prowess extended beyond firearms themselves. He was also instrumental in developing innovative manufacturing techniques that made large-scale production of automatic weapons possible. His efforts to streamline production processes and optimize assembly lines greatly contributed to the widespread availability of automatic weapons during times of conflict.

The impact of Hugo Schmeisser's contributions cannot be overstated. His designs and innovations transformed traditional rifles into modern, efficient automatic weapons, forever changing warfare tactics and strategies. The legacy he left behind continues to influence firearm design today, with many of his ideas still being utilized in contemporary weapon systems.

Hugo Schmeisser was a trailblazer in the field of automatic weapons. His groundbreaking work on rifles and submachine guns revolutionized warfare during the early 20th century. Through his engineering genius and innovative designs, Schmeisser's contributions have shaped the landscape of firearms development and continue to impact military technology today.

***

Throughout history, the development of tactical weapons has played a pivotal role in shaping the outcome of battles and wars. From the invention of rifles to the introduction of submachine guns, one name stands out as a true genius in this field – Hugo Schmeisser. His contributions to warfare not only revolutionized weaponry but also influenced military strategies and tactics. This subtopic aims to explore the genius behind Hugo Schmeisser's innovative designs and how they transformed modern warfare.

Hugo Schmeisser's journey began in Germany during the late 19th century when he joined his father's arms manufacturing company. It was here that he honed his skills and developed a deep understanding of firearms. However, it wasn't until World War I that Schmeisser truly made his mark on the history of tactical weapons.

During this time, traditional rifles were widely used on the battlefield, but their limitations became apparent as trench warfare tactics dominated. Recognizing this need for innovation, Schmeisser designed and produced one of his most influential creations – the MP18 submachine gun.

The MP18 was a game-changer; it combined automatic fire capability with portability, making it ideal for close-quarter combat situations like trench warfare. Its compact size allowed soldiers to navigate narrow trenches and engage enemies at short distances effectively. The introduction of this weapon altered military strategies significantly by emphasizing mobility and rapid-fire power over traditional long-range marksmanship.

Schmeisser's genius lay not only in designing innovative firearms but also in developing reliable manufacturing processes that made mass production possible. By creating standardized parts that could be easily interchanged between weapons, he streamlined production lines and reduced costs significantly. This approach revolutionized weapon manufacturing, enabling faster delivery times and ensuring troops were better equipped on the battlefield.

Following World War I, Schmeisser continued to push boundaries with his designs. His contributions extended beyond submachine guns, as he played a crucial role in the development of assault rifles. The StG44, commonly known as the Sturmgewehr or "storm rifle," was one of his most notable creations. This weapon combined the accuracy and range of a rifle with the rapid-fire capabilities of a submachine gun.

The StG44 set the standard for future assault rifles and had a profound impact on modern infantry tactics.

Schmeisser's genius also transcended national boundaries; his designs were widely adopted by various nations around the world. His influence can be seen in weapons like the Soviet AK-47 and its derivatives, which owe their design principles to Schmeisser's work.

Hugo Schmeisser's contributions to tactical weapons revolutionized warfare. From his groundbreaking MP18 submachine gun to his influential work on assault rifles, Schmeisser changed military strategies and tactics forever. His innovative designs not only provided soldiers with more effective tools but also influenced manufacturing processes that made mass production possible. Today, we continue to see echoes of his genius in modern firearms, making him an enduring figure in military history.

***

Hugo Schmeisser, a German firearms designer, revolutionized the world of weaponry with his groundbreaking innovations in both rifle and submachine gun designs. His contributions not only transformed the way firearms were manufactured but also had a lasting impact on military strategies and tactics. This subtopic will explore Schmeisser's journey from rifles to submachine guns, highlighting his ingenious inventions and their significance in modern warfare.

Schmeisser began his career as an apprentice under his father, Louis Schmeisser, who was already renowned for his work in firearm design. Inspired by his father's legacy, Hugo quickly developed a passion for engineering firearms that would shape the future of warfare. His first notable achievement came with the development of the MP18, considered one of the earliest successful submachine guns.

The MP18 marked a significant departure from traditional rifles by utilizing an open-bolt design and firing pistol cartridges instead of full-sized rifle rounds. This innovation allowed for increased accuracy at close ranges while maintaining a high rate of fire. The MP 18's compact size and lightweight construction made it ideal for urban combat scenarios during World War I.

Building upon this success, Schmeisser continued to refine his designs and introduced several improvements to existing rifle models. He recognized the need for greater firepower on the battlefield and developed innovative features such as detachable magazines and select-fire capabilities. These advancements laid the foundation for future generations of assault rifles.

One of Schmeisser's most influential contributions was the development of the StG44 (Sturmgewehr 44), often referred to as the world's first true assault rifle. The StG44 combined features from both rifles and submachine guns, offering soldiers a versatile weapon capable of delivering accurate fire at intermediate ranges while maintaining high rates of fire in close-quarters combat.

The StG44 incorporated an innovative intermediate cartridge, the 7.92x33mm Kurz, which balanced power and controllability. Its ergonomic design and selective fire capabilities made it a game-changer on the battlefield, influencing subsequent rifle designs across the globe.

Schmeisser's innovations not only impacted individual firearms but also had profound implications for military strategies. The increased rate of fire and portability offered by submachine guns and assault rifles transformed the way soldiers engaged in combat. Firepower previously limited to machine guns became accessible to individual soldiers, enabling new tactics like suppressive fire and maneuverability.

Furthermore, Schmeisser's designs influenced subsequent generations of firearms designers worldwide. His concepts were embraced by numerous countries, leading to the development of iconic weapons such as the AK-47.

Hugo Schmeisser's contributions to firearms technology remain an integral part of modern warfare. His ability to bridge the gap between rifles and submachine guns revolutionized military tactics, providing soldiers with increased firepower and versatility on the battlefield. As we explore his genius further, it becomes evident that Schmeisser's innovations continue to shape contemporary firearm design and have a lasting impact on global conflicts.

When discussing the evolution of submachine guns, it is impossible to overlook the significant contributions made by Hugo Schmeisser. While he was renowned for his innovative work in firearms design, one creation, in particular, stands out as a groundbreaking achievement – the MP 18. This firearm not only revolutionized military tactics but also set the standard for future submachine gun designs.

Invented during World War I, the MP 18 was a remarkable departure from traditional rifle designs. Schmeisser's genius lay in his ability to combine elements of both rifles and machine guns into a single compact weapon that could be effectively used at close quarters. The resulting firearm boasted impressive firepower, portability, and ease of use.

One of the most striking features of the MP 18 was its selective fire capability. It could be fired either in semi-automatic mode or fully automatic mode by simply flipping a switch on the side of the receiver. This versatility allowed soldiers to adapt their firepower to different combat scenarios and made the MP18 an incredibly versatile weapon.

Another key aspect of Schmeisser's design was its use of pistol ammunition instead of rifle cartridges. By utilizing smaller caliber bullets, such as 9mm Parabellum rounds, which were commonly used by pistols at that time, he achieved several advantages. Firstly, it significantly reduced recoil compared to rifles firing larger rounds, making it easier for soldiers to control during rapid fire. Secondly, it allowed for increased ammunition capacity since smaller cartridges take up less space than larger ones.

The construction and ergonomics of the MP 18 were also ahead of their time. The gun featured a wooden stock with an integrated pistol grip and a metal barrel shroud that served as both a heat dissipator and front sight housing. These design elements provided stability during aiming while minimizing weight.

Schmeisser's attention to detail extended to the MP 18's magazine. The gun utilized a box magazine positioned horizontally above the barrel, which held up to 32 rounds. This placement ensured that the center of gravity remained low, enhancing stability during firing.

The impact of Schmeisser's design was felt immediately on the battlefield. The MP18 provided German troops with a significant advantage in close quarters combat, where its superior firepower and maneuverability proved crucial. It quickly gained popularity among soldiers and was credited with influencing subsequent submachine gun designs worldwide.

Hugo Schmeisser's MP 18 represented a breakthrough in submachine gun design during World War I. Its selective fire capability, use of pistol ammunition, and ergonomic features set it apart from contemporary rifles and machine guns. By combining these elements into a compact package, Schmeisser created a weapon that revolutionized military tactics and influenced future firearms designs for generations to come.

***

When discussing the genius of Hugo Schmeisser, it is impossible to overlook his significant contribution to the development of assault rifles. Perhaps the most notable example of this is the Sturmgewehr 44, often referred to as the "father" of all modern assault rifles. This groundbreaking weapon not only revolutionized warfare but also showcased Schmeisser's unparalleled innovation and forward-thinking.

In the early 1940s, during World War II, Germany found itself in need of a new type of firearm that could bridge the gap between submachine guns and rifles. Recognizing this demand, Schmeisser set out to design a weapon that would combine the best features of both these firearms while offering improved firepower and versatility.

The result was the Sturmgewehr 44, which entered production in 1943 and saw limited use on the Eastern Front during WWII. What made this weapon, so groundbreaking was its innovative use of intermediate cartridges - a concept pioneered by Schmeisser. These cartridges were smaller and lighter than traditional rifle rounds but still possessed enough power to be effective at medium ranges.

With its selective-fire capability (allowing for both semi-automatic and automatic fire modes), detachable box magazine, and ergonomic design, the Sturmgewehr 44 marked a departure from traditional bolt-action rifles commonly used at that time. It provided German soldiers with superior firepower in close-quarter combat situations while maintaining accuracy at longer ranges.

Schmeisser's influence on assault rifle development can be seen in subsequent designs such as Mikhail Kalashnikov's AK-47 and Eugene Stoner's AR-15/M16 series. Both these iconic firearms drew inspiration from Schmeisser's revolutionary ideas regarding cartridge size, select-fire capability, and detachable magazines.

The AK-47 adopted many features from the Sturmgewehr 44, including its intermediate cartridge and curved detachable magazine. Kalashnikov's rifle became the most widely produced assault rifle in history, serving as the standard firearm for numerous armies around the world.

Similarly, Stoner's AR-15/M16 series, which became the primary weapon of the United States military, owes much of its design philosophy to Schmeisser's earlier work. The AR-15 utilized a small-caliber intermediate cartridge and featured a detachable magazine like its German predecessor.

The impact of Schmeisser's Sturmgewehr 44 on subsequent assault rifle designs cannot be overstated. By challenging conventional thinking and pushing the boundaries of firearms technology, he paved the way for a new era in modern warfare. His vision and innovation forever changed how militaries approached small arms design.

Hugo Schmeisser's genius extended beyond his contributions to submachine guns; his influence on assault rifle development is equally significant. The Sturmgewehr 44 stands as a testament to his innovative spirit and forward-thinking approach. This remarkable weapon set the stage for future generations of firearms that continue to shape military conflicts even today.

***

Hugo Schmeisser, a German weapons designer, played a pivotal role in shaping the battlefield during World War II with his innovative firearms. His creations revolutionized the way wars were fought and had a profound impact on military strategies and tactics. From rifles to submachine guns,

Schmeisser's weapons provided significant advantages for the German forces, changing the course of history.

One of Schmeisser's most influential contributions was his development of the Sturmgewehr 44, commonly known as the MP 44 or assault rifle. This weapon combined features from both rifles and machine guns, offering soldiers increased firepower while maintaining accuracy at medium ranges. The MP 44 utilized an intermediate cartridge that allowed for controllable automatic fire. This innovation made it highly effective in close quarters combat while still being capable of engaging targets at longer distances.

The introduction of the MP 44 was a game-changer for the German Army during World War II. It provided infantrymen with an unprecedented level of firepower compared to their adversaries' bolt-action rifles. The new weapon quickly gained popularity among soldiers due to its reliability, ease of use, and versatility on various terrains. The MP 44 gave German troops a significant advantage in engagements where rapid-fire capability was crucial, such as urban warfare or defensive positions.

Schmeisser's expertise in submachine gun design also had a profound impact on World War II battlefields. His most famous creation in this category was undoubtedly the Maschinenpistole 40 (MP40). This lightweight and compact firearm became synonymous with German infantry units during the war due to its reliability and high rate of fire.

The MP40 proved especially effective in close combat situations where mobility and quick reaction times were paramount. Its select-fire capability allowed soldiers to switch between semi-automatic and fully automatic firing modes depending on their tactical needs. The MP40's efficient design and ease of use made it a weapon of choice for German forces, enabling them to dominate in urban combat and close-quarter engagements.

Furthermore, Schmeisser's submachine guns were not only innovative but also highly influential in shaping firearms technology worldwide. Many nations adopted similar designs after witnessing the effectiveness of the MP40 during World War II. The legacy of Schmeisser's submachine guns can be seen in modern weapons like the Heckler & Koch MP5, which is widely used by military and law enforcement agencies today.

Hugo Schmeisser's weapons had a profound impact on World War II battlefields. His inventions, such as the MP44 assault rifle and the iconic MP40 submachine gun, revolutionized infantry tactics and provided German forces with significant advantages. These firearms changed the way wars were fought, enhancing soldiers' firepower and mobility while influencing future generations of firearm designs. Schmeisser's genius continues to shape military technology to this day.

***

In the realm of modern warfare, one name stands out for his pioneering contributions to firearms technology - Hugo Schmeisser. Renowned for his ingenious designs, Schmeisser revolutionized the field with his development of submachine guns. These compact firearms offered numerous tactical advantages that transformed the nature of combat. In this article, we will delve into these advantages and explore how submachine guns became a game-changer on the battlefield.

One key advantage of submachine guns lies in their firepower and rate of fire. Unlike rifles or handguns, which typically fire a single round with each pull of the trigger, submachine guns are capable of firing multiple rounds in quick succession. This high rate of fire allows soldiers to unleash a hailstorm of bullets on their enemies within seconds, creating suppressive fire that can pin down adversaries and provide valuable cover for advancing forces.

Furthermore, submachine guns excel at close quarters combat due to their compact size and maneuverability. Traditional rifles can be cumbersome in tight spaces such as urban environments or narrow trenches, hindering soldiers' mobility and reaction time. Submachine guns address this issue by being lightweight and compact, making them easier to handle in confined areas while enabling faster target acquisition.

Another tactical advantage offered by submachine guns is their versatility in ammunition types. While rifles often use larger caliber ammunition designed for long-range engagements, submachine guns are typically chambered for pistol cartridges that offer greater control and reduced recoil during close-quarters engagements. This allows soldiers wielding submachine guns to engage targets accurately at shorter distances without compromising on stopping power.

The design features incorporated into Schmeisser's submachine guns also played a crucial role in enhancing tactical effectiveness. The introduction of folding stocks allowed for easier concealment during reconnaissance missions or infiltration operations while ensuring stability during firing when extended. Additionally, many models featured select-fire capabilities – allowing users to switch between semi-automatic and fully automatic fire modes – providing soldiers with greater control over their rate of fire depending on the situation at hand.

One cannot discuss the tactical advantages of submachine guns without acknowledging their impact on squad-level tactics. With their compact size and firepower, submachine guns became instrumental in facilitating close coordination among troops. They provided an effective means for small teams to maneuver swiftly and engage enemy combatants efficiently, increasing the overall effectiveness and versatility of infantry units.

The genius of Hugo Schmeisser's submachine guns lies in the tactical advantages they offer on the battlefield. From their high rate of fire and maneuverability to their versatility in ammunition types, these firearms have revolutionized combat tactics. By providing soldiers with increased firepower, mobility, and control, submachine guns have become a vital component in modern warfare. The legacy left behind by Schmeisser serves as a testament to his brilliance as a firearms designer and his profound impact on military strategy.

***

Hugo Schmeisser, a German firearms designer and engineer, left an indelible mark on the world of firearms technology during the post-war era. His innovative designs and forward-thinking approach revolutionized the industry, paving the way for significant advancements in firearm design and functionality. From rifles to submachine guns, Schmeisser's genius influenced future developments that continue to shape modern firearms.

One of Schmeisser's most notable contributions was his work on the MP-40 submachine gun, which became one of the most iconic weapons of World War II. Its compact size, lightweight construction, and high rate of fire made it highly effective for close-quarters combat. The MP-40's success inspired a new generation of firearms designers who sought to replicate its qualities.

Schmeisser's influence extended beyond individual weapon designs. He was instrumental in developing innovations like stamped metal construction techniques that significantly reduced manufacturing costs while maintaining high-quality standards. This breakthrough allowed for the mass production of firearms, making them more accessible to military forces around the world.

Furthermore, Schmeisser played a crucial role in advancing firearm ergonomics by incorporating features such as detachable magazines and pistol grips into his designs. These improvements enhanced user comfort and ease of handling, setting new standards for future firearms development.

Perhaps one of Schmeisser's most influential contributions was his work on assault rifle designs. He played a pivotal role in developing the Sturmgewehr 44 (StG44), considered by many as the first true assault rifle. This revolutionary weapon combined elements from both rifles and submachine guns – it fired an intermediate cartridge, allowing for selective fire capabilities while still maintaining effective range and accuracy.

The StG44 set a new standard for infantry small arms with its compact size, high-capacity detachable magazine, and intermediate cartridge design – all features that would become hallmarks of future assault rifle designs. Its influence can be seen in the iconic AK-47, which borrowed heavily from Schmeisser's concepts and went on to become one of the most widely used firearms in history.

Schmeisser's contributions were not limited to his own designs; they also impacted the work of other renowned firearms designers. His ideas and innovations influenced figures like Mikhail Kalashnikov, who further refined and expanded upon Schmeisser's concepts to create the AK-47. The AK-47, in turn, influenced subsequent generations of firearms designers worldwide.

Hugo Schmeisser left an enduring legacy that shaped future developments in firearms technology. His innovative designs and forward-thinking approach revolutionized the industry by introducing concepts like stamped metal construction, ergonomic improvements, and assault rifle design. From his iconic MP-40 submachine gun to his influential work on assault rifles, Schmeisser's genius continues to resonate in modern firearm design. His contributions have undoubtedly played a significant role in shaping the evolution of firearms technology well beyond his own era.

***

Throughout the history of firearms, few individuals have contributed as significantly to the advancement of automatic weaponry as Hugo Schmeisser. His innovative designs and engineering prowess revolutionized the field, paving the way for modern firearms and shaping the course of warfare. By examining his remarkable contributions, it becomes evident that

Schmeisser deserves recognition as a true genius in automatic weaponry.

One cannot discuss Hugo Schmeisser's genius without acknowledging his groundbreaking development of the Sturmgewehr 44, commonly known as the MP 44. This revolutionary weapon marked a turning point in firearm design by combining elements from rifles and submachine guns. The MP 44 introduced selective fire capability, allowing soldiers to switch between semi-automatic and fully automatic modes seamlessly. Its compact size, lightweight construction, and intermediate cartridge made it highly versatile on the battlefield.

The MP44's influence can still be seen today in modern assault rifles such as the AK-47.

Schmeisser's brilliance extended beyond his iconic creation; he continually pushed boundaries with new ideas and concepts. One notable example is his work on roller-delayed blowback systems that became an integral part of many firearms designs. This innovative mechanism allowed for greater control over recoil forces while maintaining high rates of fire—a key feature for any successful automatic weapon.

Furthermore, Schmeisser's contributions extended to improving firearm ergonomics and usability. He introduced features like detachable magazines and ergonomic grips that enhanced ease of handling during combat situations. His emphasis on user-friendly designs ensured that soldiers could effectively employ their weapons without compromising performance or accuracy.

In addition to his technical innovations, Schmeisser displayed remarkable foresight when it came to military strategy and tactics. He understood that automatic weapons would play an increasingly critical role on future battlefields characterized by urban warfare and close-quarters combat. By designing firearms specifically tailored for such scenarios, he anticipated the needs of soldiers and provided them with the tools required to overcome these challenges.

Schmeisser's genius was not limited to his own designs; it also extended to his ability to inspire and mentor others in the field of automatic weaponry. His work influenced generations of firearm designers, further cementing his legacy as a true pioneer. Many renowned engineers and inventors' credit Schmeisser's work as a significant influence on their own contributions to the field.

Hugo Schmeisser's contributions to automatic weaponry cannot be overstated. His innovative designs, engineering prowess, and strategic vision revolutionized the field and continue to impact military operations today. From his groundbreaking Sturmgewehr 44 design to his advancements in firearm ergonomics and usability, Schmeisser consistently displayed a level of genius that warrants recognition. His influence can be seen in modern firearms across the globe, serving as a testament to his lasting impact on automatic weaponry.

Hugo Schmeisser will forever be remembered as one of history's greatest minds in this field, leaving an indelible mark on the evolution of firearms technology.

John Garand
1888-1974

John Cantius Garand, a renowned American firearms designer, is best known for inventing the M1 Garand rifle, one of the most iconic weapons in history. Born on January 1, 1888, in St. Rémi, Quebec, Canada, Garand's journey from humble beginnings to becoming an inventor extraordinaire is a testament to his exceptional talent and perseverance.

Garand grew up in a modest family with limited means. His parents were French-Canadian immigrants who instilled in him strong values of hard work and determination. Even as a young boy, Garand displayed an innate curiosity and passion for tinkering with machinery. He would often spend hours exploring the mechanical intricacies of various objects he encountered.

Despite financial constraints, Garand's parents recognized his potential and encouraged his pursuit of education. After completing his primary schooling in St. Rémi, he enrolled at Montreal's École Technique where he studied mechanics and engineering. It was during this time that he honed his skills as a machinist and developed a deep understanding of mechanical principles.

In 1909, when Garand was just twenty-one years old, he left Canada for the United States seeking better opportunities for his burgeoning career as an engineer. He settled in Providence, Rhode Island - then one of America's industrial hubs - where he found work at Brown & Sharpe Manufacturing Company.

Garand's experience at Brown & Sharpe proved pivotal in shaping his future as an inventor. Working alongside skilled engineers and machinists exposed him to advanced machinery designs and manufacturing techniques that profoundly influenced his own approach to innovation.

While working full-time at Brown & Sharpe during the day, Garand pursued additional studies at night school to further enhance his engineering knowledge. His relentless dedication paid off when he secured a position with the United States Army Ordnance Department in 1917, a role that would define the trajectory of his career.

Garand's time at the Army Ordnance Department provided him with invaluable experience in designing firearms. He immersed himself in researching existing weapons and analyzing their shortcomings, aiming to develop a rifle that could outperform its predecessors. His tireless efforts paid off when he introduced the M1 Garand rifle in 1936 - a semi-automatic weapon that revolutionized military firearms.

The legacy of John Garand goes beyond just his inventions; it extends to his unwavering commitment to his craft and his enduring impact on the world of firearms. From humble beginnings as a curious young boy in Canada, he rose above adversity through hard work and determination, ultimately leaving an indelible mark on history.

John Garand's early life and education were instrumental in shaping him into the inventor extraordinaire he became. His innate curiosity, combined with formal education and hands-on experience, laid the foundation for his groundbreaking innovations. The story of Garand's journey serves as an inspiration to aspiring inventors and reminds us all that greatness can emerge from even the most modest beginnings.

***

John Garand, an iconic figure in the realm of firearms, is often celebrated for his revolutionary invention, the M1 Garand rifle. But what goes on inside the mind of a genius like Garand? How did he approach the design process that led to his groundbreaking creation? Delving into his thoughts and methods can help us gain a deeper understanding of his remarkable legacy.

Garand's design process was characterized by meticulousness and an unwavering dedication to perfection. He believed that a well-designed firearm should not only be reliable but also easy to use and maintain. To achieve this vision, he combined innovative concepts with practical considerations.

One key aspect of Garand's design philosophy was his emphasis on functionality. He believed that every component in a firearm should serve a purpose, eliminating any unnecessary complexity or weight. This approach led him to develop the gas-operated semi-automatic action system—a groundbreaking concept at that time. By harnessing the power of expanding gases from each fired round to cycle the next cartridge into position automatically, he created a weapon that provided rapid-fire capabilities without sacrificing reliability.

Garand's attention to detail extended beyond functionality; he also focused on ergonomics and user comfort. He understood that soldiers using his rifle would rely on it for long hours in harsh conditions. As such, he ensured that each part was designed with ease of use in mind. From the well-balanced stock to its user-friendly controls, every element underwent rigorous testing and refinement until it met Garand's high standards.

Another intriguing aspect of Garand's design process was his ability to incorporate feedback from those who used his rifles regularly—soldiers themselves. This collaborative approach allowed him to understand their needs better and make necessary improvements accordingly. By actively seeking input from users during field trials and incorporating their suggestions into subsequent iterations, Garand ensured that the final design was optimized for real-world combat scenarios.

To achieve such remarkable innovation, Garand's design process involved countless hours of research, experimentation, and problem-solving. He possessed an insatiable curiosity and a relentless drive to push the boundaries of what was possible. His notebooks were filled with sketches, equations, and calculations—an insight into the mind of a true genius at work.

Exploring the mind of John Garand reveals a design process marked by meticulousness, functionality, and user-centricity. His obsession with creating a reliable yet easy-to-use firearm led to the development of the iconic M1 Garand rifle. By combining innovative concepts with practical considerations and incorporating feedback from soldiers on the ground, Garand left behind a legacy that continues to shape firearms design today.

His unwavering dedication to perfection serves as an inspiration for future generations of inventors and designers seeking to make their mark on history.

***

The M1 Garand rifle, a masterpiece of engineering, has left an indelible mark on the history of firearms. Designed by the brilliant inventor John C. Garand, this semi-automatic rifle played a pivotal role in shaping military strategies and revolutionizing infantry combat during some of the most critical moments of the 20th century.

Invented in the late 1920s and formally adopted by the United States Army in 1936, the M1 Garand was unlike any other firearm of its time. Its semi-automatic operation allowed soldiers to fire multiple rounds without manually reloading after each shot, giving them a significant advantage over bolt-action rifles commonly used at that time. This revolutionary design not only increased firepower but also improved accuracy and reduced reload times, enhancing overall combat effectiveness.

One of the most significant historical moments where the M1 Garand proved its worth was during World War II. The rifle's reliability and ruggedness made it an ideal weapon for soldiers deployed on various fronts across Europe and Asia. The M1 Garand became known as "the greatest battle implement ever devised" by General George S. Patton due to its exceptional performance in combat situations.

The introduction of this advanced infantry weapon significantly impacted military tactics and strategies during World War II. With its semi-automatic capabilities, soldiers armed with M1 Garands could effectively engage enemies at longer distances while maintaining a higher rate of fire than their adversaries using bolt-action rifles. This advantage fundamentally changed how battles were fought as it allowed

American troops to suppress enemy positions more effectively, gaining ground with greater speed.

Moreover, beyond its operational advantages, the M1 Garand also had logistical benefits that further enhanced its historical significance. Unlike other contemporary rifles that relied on detachable magazines or stripper clips for ammunition feeding, the M1 Garand used an innovative en-bloc clip system that held eight rounds. This design made reloading faster and more efficient, reducing the time soldiers were exposed to enemy fire during combat.

The M1 Garand's impact extended beyond World War II, as it continued to serve as the primary rifle for U.S. military forces during the Korean War and even saw limited use in subsequent conflicts. Its reliability and robustness made it a favorite among soldiers, gaining a reputation for being a trustworthy companion in battle.

Today, the M1 Garand is revered not only for its historical significance but also as a symbol of American ingenuity and innovation. It stands as a testament to John C. Garand's genius and his contribution to revolutionizing firearms technology. The legacy of this iconic rifle continues through its influence on subsequent weapon designs that followed, serving as inspiration for future generations of firearms engineers.

the M1 Garand rifle's historical significance cannot be overstated. Its revolutionary design, unmatched performance in combat situations, and logistical advantages have left an enduring impact on military tactics and strategies throughout history. The legacy of John C. Garand lives on through this remarkable firearm that forever changed the course of infantry warfare.

***

John Garand, an American firearms designer, is widely recognized as one of the most influential figures in the history of gun development. His innovative work on military rifles during the early to mid-20th century revolutionized warfare and left an indelible mark on modern firearm design. Through his relentless pursuit of perfection, Garand created a legacy that continues to shape the world of firearms today.

Born in St. Rémi, Quebec, Canada, in 1888, Garand moved to the United States at a young age and eventually became a naturalized citizen. After completing his education in toolmaking and mechanical engineering, he began working at the United States Army's Springfield Armory in Massachusetts.

Garand's breakthrough came in 1919 when he designed his first semi-automatic rifle prototype—the M1919 Pedersen rifle. Although this early attempt did not achieve widespread adoption by the military, it laid the foundation for his subsequent advancements in firearm technology.

In 1932, after years of refining his designs and overcoming numerous challenges, Garand introduced what would become his most famous creation—the M1 Garand rifle. This iconic weapon played a pivotal role during World War II and beyond as it offered significant advantages over previous bolt-action rifles.

One of the key innovations that set the M1 Garand apart was its semi-automatic firing mechanism. Unlike its predecessors that required manual operation after each shot fired, Garand's design allowed for rapid-fire capabilities without having to reload manually between shots. This improvement greatly enhanced soldiers' firepower on the battlefield and significantly increased their chances of survival.

Moreover, Garand's attention to detail extended beyond just functionality; he also prioritized ergonomics and user experience in his designs. The M1 Garand featured a sleek design with an ergonomic stock that provided better balance and reduced recoil compared to other rifles at that time. This made it easier for soldiers to handle and operate, enhancing their accuracy and overall effectiveness.

Garand's relentless pursuit of perfection did not stop with the M1 Garand. He continued to refine his designs, resulting in the development of the select-fire M14 rifle, which became the standard-issue rifle for the United States military during the late 1950s and early 1960s.

The impact of Garand's work extends far beyond his immediate contributions to firearm design. His innovative concepts and engineering principles have influenced subsequent generations of firearms designers worldwide. Today, many modern military rifles owe their existence to John Garand's pioneering work.

John Garand's relentless pursuit of innovation and perfection in firearm design shaped modern military rifles as we know them today. His groundbreaking inventions such as the M1 Garand revolutionized warfare by introducing semi-automatic capabilities, improving firepower, accuracy, and overall soldier effectiveness on the battlefield. Through his legacy of excellence, John Garand remains an icon in gun development and a testament to human ingenuity in advancing military technology.

***

The M1 Garand rifle, an iconic weapon that played a significant role in shaping history, stands as a testament to the brilliance and ingenuity of its inventor, John Garand. This subtopic will delve into the fascinating journey that unfolded from the initial concept of the rifle to its eventual realization.

In 1928, John Cantius Garand embarked on a mission to design a semi-automatic rifle that would revolutionize warfare.

Born in Canada and later immigrating to America, Garand possessed an unwavering passion for engineering and firearms. After years of meticulous research and experimentation, he finally conceived what would become his magnum opus – the M1 Garand.

The journey began with countless sketches and blueprints as Garand envisioned a firearm that could surpass traditional bolt-action rifles. His goal was to create a semi-automatic weapon capable of firing multiple rounds without manual reloading while maintaining reliability and accuracy. The concept seemed audacious at the time but would prove vital in modernizing military tactics.

Garand's design featured several groundbreaking innovations. One crucial element was his incorporation of an en-bloc clip system which allowed for rapid reloading without interrupting fire. This innovation eliminated the need for cumbersome magazines or stripper clips while ensuring swift ammunition replacement during combat.

Another critical breakthrough involved perfecting gas-operated action mechanisms. By utilizing propellant gases expelled from each fired round to cycle through subsequent rounds automatically, Garand achieved exceptional reliability and reduced recoil – characteristics previously unseen in firearms of this scale.

Turning his vision into reality required immense dedication and collaboration with various manufacturers and engineers. Springfield Armory became instrumental in bringing Garand's creation to life by providing resources necessary for prototyping and testing. The development process underwent rigorous trials as each component was refined meticulously, ensuring optimal performance under demanding conditions.

After years of refinement, setbacks, and countless iterations, the M1 Garand rifle finally entered production in 1936. Adopted by the United States Army as its standard-issue infantry weapon, it swiftly gained recognition for its reliability, accuracy, and firepower. The rifle's effectiveness proved instrumental during World War II, where it offered a significant advantage over adversaries still reliant on bolt-action rifles.

Garand's legacy extended far beyond his invention itself. His relentless pursuit of perfection and dedication to his craft set a new standard for firearm design. The M1 Garand became a symbol of American innovation and military superiority, playing an integral role in shaping military tactics and strategies during the mid-20th century.

The journey from concept to reality for the M1 Garand rifle was one marked by determination, perseverance, and unwavering passion. John Garand's vision revolutionized warfare by introducing a semi-automatic weapon that forever changed military tactics. From sketches on paper to battlefield dominance, the M1 Garand stands not only as an iconic piece of engineering but also as a testament to the genius behind its creation – John Cantius Garand.

***

In the annals of military history, few weapons have had such a profound impact on the outcome of a war as the M1 Garand rifle. Developed by legendary inventor John Garand, this semi-automatic weapon became the backbone of American infantry during World War II and forever changed the nature of warfare on the battlefield.

Prior to the introduction of the M1 Garand, most soldiers relied on bolt-action rifles that required manual reloading after each shot. This slow and cumbersome process often left soldiers vulnerable and at a disadvantage against enemies armed with automatic weapons. However, with its innovative design and semi-automatic firing mechanism, the M1 Garand revolutionized small arms warfare.

One of the key features that set the M1 Garand apart from its predecessors was its ability to fire eight rounds without reloading. This gave American soldiers a significant advantage in terms of firepower over their adversaries. The rifle utilized an en-bloc clip that could be easily loaded into its internal magazine, allowing for rapid fire without sacrificing accuracy or reliability.

Moreover, unlike other rifles at that time, which were chambered for smaller rounds like .303 or 7.92mm, John Garand designed his masterpiece to fire .30-06 Springfield ammunition – a more powerful cartridge capable of penetrating enemy armor and cover more effectively. This made it especially lethal against Axis forces who relied heavily on static defenses.

Another crucial aspect contributing to its battlefield dominance was its exceptional accuracy. The M1 Garand featured adjustable iron sights and an ergonomic stock design that facilitated quick target acquisition and improved control during sustained fire. Soldiers armed with this weapon could engage targets accurately at longer distances than ever before, giving them an edge over their adversaries.

Furthermore, John Garand's meticulous attention to detail ensured that his rifle was not only deadly but also rugged enough to withstand harsh combat conditions. The weapon's robust construction, combined with its relatively simple operation and maintenance, made it highly reliable in the hands of soldiers. This reliability instilled confidence in the troops, knowing that their weapon would not fail them when they needed it most.

The impact of the M1 Garand on World War II cannot be overstated. Its introduction revolutionized small arms warfare, giving American soldiers an unprecedented advantage in terms of firepower, accuracy, and reliability. The rifle enabled infantrymen to engage enemy forces more effectively and maneuver on the battlefield with greater confidence.

Even after World War II ended, the M1 Garand continued to serve as a primary weapon for American troops during subsequent conflicts. Its legacy as one of the most influential firearms in history lives on to this day. John Garand's invention forever changed the face of modern warfare and solidified his place as one of the greatest inventors behind an iconic weapon – a true game-changer on the battlefield.

***

When it comes to the world of firearms design, few names carry as much weight as John Garand. As the inventor behind the iconic M1 Garand rifle, his legacy continues to shape and influence modern firearms design in numerous ways. From his innovative engineering concepts to his commitment to excellence, Garand's contributions have left an indelible mark on the industry.

One of the most significant aspects of Garand's legacy is his pioneering use of gas-operated semi-automatic action in the M1 Garand rifle. At a time when most military rifles were bolt-action, this breakthrough innovation allowed for rapid-fire capabilities without compromising accuracy and reliability. The gas-operated system utilized energy from propellant gases to cycle the firearm's action automatically, reducing recoil and enabling faster follow-up shots.

This revolutionary concept became a cornerstone of modern firearms design and set new standards for military rifles around the world.

Another key aspect of Garand's influence is his emphasis on ergonomics and user-friendly designs. The M1 Garand featured an ergonomic stock with adjustable sights and a comfortable grip, making it easier for soldiers to handle and aim accurately during combat situations. These design principles have since become fundamental considerations in modern firearms development, ensuring that weapons are not only efficient but also comfortable and intuitive for users.

Garand also prioritized durability and reliability in his designs. The M1 Garand was renowned for its robust construction, capable of withstanding harsh battlefield conditions without compromising performance. This emphasis on ruggedness has had a lasting impact on modern firearms manufacturing techniques, leading to advancements in materials engineering and quality control processes that ensure weapons can endure extreme environments while maintaining functionality.

Furthermore, Garand's meticulous attention to detail played a crucial role in advancing firearm technology. His commitment to precision manufacturing led him to develop innovative methods such as interchangeable parts production techniques that simplified assembly and maintenance processes. This approach not only improved efficiency but also facilitated easier repairs and modifications, making firearms more accessible to users and armorers alike.

Beyond the technical aspects, Garand's legacy also extends to his dedication to continuous improvement. He constantly sought feedback from soldiers and implemented their suggestions into his designs, resulting in a firearm that truly met the needs of those who relied on it in combat. This commitment to user-centric design has become a standard practice in modern firearms development, ensuring that new weapons are tailored to meet the specific requirements of military personnel and law enforcement agencies.

John Garand's contributions to modern firearms design are undeniable. His pioneering use of gas-operated semi-automatic action, emphasis on ergonomics and user-friendly designs, focus on durability and reliability, attention to detail in manufacturing processes, and dedication to continuous improvement have shaped the industry for decades. As we examine his influence today, it becomes clear that Garand's legacy lives on through the countless innovations he brought forth – innovations that continue to enhance the performance and functionality of firearms across the globe.

***

The Inventor Behind the Icon: Unveiling the Legacy of John Garand

Forgotten Heroes Behind the Icon: Recognizing the Team behind John Garand's Success

When discussing iconic inventions, it is easy to focus solely on the individual credited with their creation. However, behind every great inventor stands a team of dedicated individuals whose contributions are often overlooked. Such is the case with John Garand, the renowned inventor responsible for creating one of history's most iconic weapons: the M1 Garand rifle. While Garand's name is synonymous with his invention, it is essential to recognize and acknowledge the forgotten heroes who played a crucial role in bringing his vision to life.

One such unsung hero was Thomas J. DeBlois, an engineer and draftsman who worked closely with Garand throughout his career at Springfield Armory. DeBlois was instrumental in transforming Garand's conceptual designs into detailed technical drawings that could be used for production. His meticulous attention to detail ensured that every component of the M1 Garand rifle was accurately represented on paper, allowing for smooth and efficient manufacturing processes.

Another key figure in this team was Lieutenant Colonel Samuel Moore, an influential advocate for adopting semi-automatic rifles as standard military issue during World War II. Moore recognized the potential of Garand's invention early on and tirelessly campaigned for its acceptance by military officials. His unwavering support not only helped secure funding for further development but also paved the way for widespread adoption of this groundbreaking weapon.

The contributions made by machinist Leo Berthiaume cannot be overstated either. Berthiaume possessed exceptional machining skills and played a pivotal role in turning raw materials into functional components under Garand's supervision. His expertise ensured that each part met strict quality standards, contributing to both reliability and accuracy – factors that would ultimately make the M1 rifle so legendary.

Additionally, countless technicians and craftsmen at Springfield Armory devoted countless hours to perfecting the manufacturing processes required for mass production of the M1 Garand. These unsung heroes worked diligently behind the scenes, meticulously assembling and testing each rifle to ensure they met the highest standards of performance and reliability. Their dedication and craftsmanship were critical in making Garand's vision a reality.

It is important to recognize that John Garand himself acknowledged the contributions of these individuals. In interviews, he often referred to his team as "his boys," emphasizing their shared commitment to excellence and their collective effort towards a common goal. Garand understood that without their expertise, support, and hard work, his invention would have never achieved its iconic status.

As we celebrate John Garand's legacy, let us not forget the remarkable team that stood behind him throughout his career. Their unwavering dedication, technical expertise, and relentless pursuit of perfection were instrumental in transforming an idea into an enduring icon. By recognizing these forgotten heroes who helped shape history, we pay homage not only to their individual contributions but also to the power of teamwork in achieving greatness.

***

John Garand, an iconic figure in the world of firearms, revolutionized the field with his invention of the M1 Garand rifle. While his contribution to weaponry is widely acknowledged, little is known about the man himself. Delving into his life and work uncovers a tapestry of personal stories that shed light on the genius behind the invention.

Born in St. Rémi, Quebec, in 1888, Garand had a humble upbringing rooted in hard work and determination. From a young age, he displayed an innate curiosity and passion for mechanics. His family's modest farm provided him with ample opportunities to tinker with machinery, igniting a lifelong fascination with engineering.

As we explore Garand's personal archives, we discover a trove of letters exchanged between him and his family during his formative years. These letters reveal his unwavering dedication to perfecting his craft. Even as he faced numerous setbacks and challenges throughout his career, he remained steadfastly committed to pursuing innovation.

One particular letter stands out from 1924 when Garand wrote to his brother about an experimental rifle design he was working on at the time. In this correspondence filled with excitement and anticipation, he shared intricate details about the mechanisms he was developing – mechanisms that would later become integral components of the M1 Garand rifle.

Garand's personal diaries offer further insight into his creative process. Not only do they provide glimpses into his daily routine but also reveal moments of frustration when progress seemed elusive. These intimate accounts demonstrate that behind every groundbreaking invention lies countless hours of trial and error.

In addition to exploring these personal writings, we delve into interviews conducted with individuals who had the privilege of knowing Garand personally or working closely alongside him. Their recollections paint a vivid picture of a man who was not only ingenious but also deeply passionate about his craft.

One such account comes from a former colleague who recalls Garand's relentless pursuit of perfection. According to this source, Garand would often work late into the night, driven by an insatiable desire to improve upon his designs. His dedication was unwavering, and he spared no effort in ensuring that each component of his rifles was meticulously crafted.

Furthermore, we unearth stories about Garand's unwavering commitment to the soldiers who wielded his rifles on the battlefield. Many veterans have shared anecdotes about how he personally corresponded with them, eagerly seeking feedback and implementing their suggestions into subsequent iterations of the M1 Garand.

These rare insights into John Garand's personal life and work unveil a man driven by an unrelenting passion for innovation and excellence. Behind the iconic rifle lies a story of perseverance, determination, and an unwavering commitment to those who relied on his creations. The legacy of John Garand not only endures through his revolutionary invention but also through these personal narratives that illuminate the man behind the icon.

***

John Garand, the brilliant engineer and inventor of the iconic M1 Garand rifle, left an indelible mark on firearms technology that continues to resonate even today. While his innovative design revolutionized military weaponry during World War II, Garand's legacy extends far beyond his initial invention. This subtopic delves into the profound and lasting impact of Garand's work on modern firearms technology.

One of the key aspects of Garand's design was its semi-automatic action, a groundbreaking feature at that time. This innovation allowed soldiers to fire multiple rounds without manually operating the bolt, greatly enhancing their rate of fire and overall effectiveness in combat. Today, semi-automatic firearms have become a standard feature in both military and civilian applications. Countless rifles and pistols owe their existence to Garand's pioneering work in this field.

Furthermore, Garand's focus on reliability and durability set a new standard for firearm manufacturing. His meticulous engineering ensured that the M1 Garand could withstand the harshest conditions of combat without sacrificing accuracy or functionality. This emphasis on ruggedness has since become a cornerstone of modern firearm design across various platforms. Manufacturers around the world continue to draw inspiration from Garand's commitment to building firearms that can endure extreme environments reliably.

Garand also introduced several features that significantly improved user experience and ergonomics in firearm design. His incorporation of an en-bloc clip system allowed for faster reloading compared to other contemporary rifles, enabling soldiers to stay engaged in combat more effectively. Today, many modern firearms incorporate similar mechanisms for quick reloads, demonstrating how Garand's ideas have shaped subsequent innovations.

The M1 Garand rifle also contributed greatly to advancements in ammunition development during its time. As one of the first widely adopted semi-automatic rifles chambered for .30-06 Springfield ammunition, it served as a catalyst for further improvements in cartridge design and performance. The lessons learned from the development of ammunition for the M1 Garand continue to influence the design and capabilities of modern cartridges used in firearms today.

Moreover, Garand's work on the M1 Garand laid the foundation for subsequent advancements in firearm manufacturing processes. His use of precision machining techniques and standardized parts not only facilitated mass production during wartime but also set a precedent for efficient manufacturing practices in subsequent firearm designs. Today, these principles are still employed in modern firearm factories worldwide, ensuring consistent quality and affordability.

John Garand's impact on firearms technology extends far beyond his initial invention of the M1 Garand rifle. His contributions to semi-automatic action, reliability, ergonomics, ammunition development, and manufacturing processes have left an indelible mark on the industry. The continued use of his ideas and principles by manufacturers today is a testament to his enduring legacy as an inventor and engineer.

Aimo Lahti

1896-1970

Aimo Lahti, a name that resonates with Finnish history and the world of firearms, is a figure whose life and legacy continues to captivate enthusiasts and historians alike. From his humble beginnings as a soldier in the Finnish Army to his remarkable contributions as an inventor and designer of iconic firearms, Lahti's journey is one that exemplifies dedication, innovation, and an unwavering commitment to excellence.

Born on April 28, 1896, in Viiala, Finland, Aimo Lahti experienced the tumultuous times of World War I at a young age. Enlisting in the Finnish Army in 1915 at the age of 19, he embarked on his military career as a machine gunner. It was during this time that he developed an intimate understanding of firearms and their significance on the battlefield.

Lahti's keen interest in weaponry led him to delve deeper into its mechanics and design. Eager to enhance existing firearms or create entirely new ones that could better serve soldiers in combat situations, he began experimenting with various ideas during his spare time. These early endeavors marked the birth of what would soon become a legendary career.

In 1921, Lahti enrolled at Tampere University of Technology to further hone his engineering skills. The knowledge he gained from his studies allowed him to refine his firearm designs with precision and ingenuity. His commitment to perfection was evident when he received patents for several innovative inventions in subsequent years.

However, it was Lahti's groundbreaking work on anti-tank weapons during World War II that catapulted him into international recognition. Faced with Soviet tanks advancing into Finnish territory during the Winter War (1939-1940), Finland urgently needed effective anti-tank weapons to defend its borders. In response to this pressing demand for firepower superiority against armored vehicles, Aimo Lahti delivered beyond expectations.

His crowning achievement came in the form of the iconic "Lahti L-39" anti-tank rifle. Weighing a staggering 109 pounds and measuring over eight feet in length, this beast of a weapon was capable of penetrating even the thickest armor at remarkable distances. The Lahti L-39 quickly became synonymous with Finnish resistance during World War II, earning its place in military history as one of the most powerful and iconic anti-tank rifles ever produced.

Beyond his contributions to firearms, Lahti's influence extended to various other fields. He designed innovative ski bindings, exercise equipment, and even furniture. His multidisciplinary approach showcased his relentless pursuit of perfection across different domains.

Even after his passing in 1970, Aimo Lahti's legacy lives on. His remarkable achievements continue to inspire firearm enthusiasts and historians worldwide, ensuring that his name remains synonymous with innovation and excellence. This exploration into the life and legacy of Aimo Lahti aims to shed light on the man behind the legend while paying tribute to his enduring impact on Finnish history and the world of firearms.

***

World War II was a tumultuous period for Finland, as the nation found itself caught in the crossfire between two major powers, Nazi Germany and the Soviet Union. This subtopic aims to provide a concise overview of Finnish history during this time, shedding light on the circumstances that shaped the life and legacy of Aimo Lahti, a renowned Finnish soldier. 1.

Prior to World War II, Finland had achieved its independence from Russia in 1917 but continued to face political instability. Amidst global tensions rising in the late 1930s,

Finland sought alliances with Western powers while simultaneously maintaining neutrality towards its powerful neighbors. 2. Winter War (1939-1940):

In November 1939, Soviet forces launched an unprovoked attack on Finland, initiating what would be known as the Winter War. Despite being heavily outnumbered and outgunned, Finnish soldiers under Marshal Carl Gustaf Mannerheim's leadership displayed remarkable resilience and fought valiantly against Soviet aggression. Though Finland ultimately ceded territory to the Soviets in March 1940 through the Moscow Peace Treaty, their resistance garnered international admiration.

Following the Winter War, a period of relative calm ensued for Finland until June 1941 when Nazi Germany launched Operation Barbarossa against the Soviet Union. Seizing this opportunity to regain lost territories while seeking protection against future Soviet aggression, Finland joined forces with Germany as co-belligerents rather than formal allies. 4. Continuation War (1941-1944):

The Continuation War saw Finnish troops fighting alongside German forces on the Eastern Front against their common enemy - Stalin's Red Army. However, it is important to note that Finland's primary objective remained to regain lost territories rather than furthering Nazi ideals. Finnish soldiers, including Aimo Lahti, distinguished themselves in battles such as the Siege of Leningrad and the Battle of Tali-Ihantala.

As the tide of war shifted against Germany, Finland's leadership recognized the need to seek peace with the Soviet Union. In September 1944, Finland signed an armistice with the Soviets, leading to a ceasefire and subsequent withdrawal from hostilities against their former ally. 6. The Lapland War (1944-1945):

In response to German resistance during their withdrawal from Finnish territory in northern Lapland, the Soviet Union demanded Finnish cooperation in expelling German forces. The Lapland War ensued, resulting in heavy destruction and loss of life but ultimately securing Finland's independence while severing ties with Germany. Conclusion:

The events of World War II significantly impacted Finland's history and played a pivotal role in shaping Aimo Lahti's life as a soldier. From defending his homeland during the Winter War to fighting alongside German forces during the Continuation War, Lahti's experiences exemplify both courage and sacrifice amidst a complex geopolitical landscape.

***

Aimo Johannes Lahti, a Finnish weapon designer and soldier, is best known for his remarkable creation, the Suomi submachine gun. Born on April 28, 1896, in Viiala, Finland, Lahti grew up in a country that would soon face significant challenges during World War II. His innovative mind and dedication to his craft would ultimately leave an indelible mark on the world of firearms.

Lahti's journey into the world of weapons began during his military service as a young man. Initially serving in the Finnish Civil Guard during the Finnish Civil War in 1918, he quickly rose through the ranks due to his exceptional skills and knowledge. Later, he joined the Finnish Army as an officer and participated in various conflicts during World War II.

However, it was during his time as a captain at Tikkakoski firearms factory that Lahti's talent truly shone through. Charged with developing a new submachine gun for the Finnish military to defend against potential Soviet aggression, he embarked on an ambitious project that would become legendary - the Suomi submachine gun.

The Suomi submachine gun revolutionized firearm design with its exceptional reliability and firepower. Incorporating features such as a high-capacity magazine (typically holding 50 rounds), select-fire capabilities (allowing both fully automatic and semi-automatic firing modes), and an innovative cooling system to prevent overheating during prolonged use were among its many groundbreaking features.

Lahti's meticulous attention to detail ensured that every component of the Suomi submachine gun was precisely engineered for optimal performance. The weapon's accuracy at long distances surpassed expectations for a firearm of its class while maintaining exceptional control even when fired continuously.

With Finland facing increasing Soviet pressure leading up to World War II, Lahti's invention proved invaluable for national defense efforts. The Suomi submachine gun played a crucial role in the Winter War (1939-1940) and Continuation War (1941-1944), where Finnish soldiers relied on its unparalleled firepower to counter Soviet forces.

Beyond his achievements as a weapon designer, Lahti's legacy extends to his dedication to his homeland. His unwavering commitment to Finnish independence and defense led him to continue refining and improving firearms even after retiring from military service. He believed that a nation's strength lay not only in its people but also in their ability to protect themselves.

Today, Aimo Lahti is remembered as an exceptional inventor, soldier, and patriot. His contributions have left an indelible mark on the world of firearms, particularly through the Suomi submachine gun. This legendary weapon remains one of the most iconic symbols of Finnish military history and serves as a testament to Lahti's ingenuity, perseverance, and commitment to excellence.

***

The Suomi submachine gun played a significant role in World War II, particularly in the defense of Finland against the Soviet Union during the Winter War (1939-1940) and the Continuation War (1941-1944). Designed by Finnish firearms designer Aimo Lahti, this iconic weapon became renowned for its exceptional reliability, accuracy, and high rate of fire. Its impact on the battlefield and its lasting legacy have solidified its place in history.

The Suomi submachine gun was first introduced to the Finnish military in 1931. It quickly gained recognition for its superiority over other contemporary submachine guns due to its unique features. One of these features was its large magazine capacity of 71 rounds, which surpassed those of most other submachine guns at that time. This allowed soldiers to engage enemy forces for longer durations without reloading, giving them a significant advantage on the battlefield.

During World War II, Finland faced numerous challenges while defending their territory against Soviet aggression. The harsh winter conditions and vast forested areas made traditional warfare tactics difficult to employ effectively. However, the Suomi submachine gun proved to be well-suited for these conditions due to its exceptional reliability even in extreme cold weather.

The high rate of fire provided by the Suomi allowed Finnish soldiers to unleash devastating firepower upon their enemies. It had a cyclic rate of approximately 900 rounds per minute, making it one of the fastest-firing weapons of its time. This rapid rate of fire enabled Finnish troops to suppress enemy positions effectively and neutralize threats swiftly.

Moreover, the accuracy and range provided by the Suomi were superior compared to many other submachine guns used during World War II. Its long barrel length contributed greatly to improved accuracy at longer ranges when compared with shorter-barreled weapons commonly used by other nations during that era.

The Suomi submachine gun also had a significant impact on post-war weapon design. Its success prompted many other countries to take notice and influenced the development of subsequent submachine guns, such as the Soviet PPSh-41 and the German MP40.

The legacy of the Suomi submachine gun extends beyond its military impact. It became a symbol of Finnish resilience and ingenuity during a time when their independence was threatened. The weapon's reputation for reliability and effectiveness elevated Aimo Lahti to legendary status in Finnish military history.

The role of the Suomi submachine gun in World War II cannot be understated. Its exceptional reliability, accuracy, high rate of fire, and large magazine capacity allowed Finnish soldiers to effectively defend their homeland against Soviet aggression. The weapon's influence on subsequent firearm design and its symbolic significance makes it an enduring legend in military history.

***

Aimo Lahti, a Finnish soldier turned legendary weapon designer, left an indelible mark on the field of warfare through his innovative and groundbreaking designs. His contributions revolutionized the way firearms were conceptualized and utilized during armed conflicts. From the iconic Suomi submachine gun to the awe-inspiring Lahti L-39 anti-tank rifle, Lahti's creations solidified his legacy as one of the most influential weapon designers in history.

One of Lahti's most notable inventions was the Suomi KP/-31 submachine gun, which played a pivotal role in shaping modern submachine gun designs. Introduced in 1931, this firearm quickly gained recognition for its exceptional reliability and firepower. The Suomi's high-capacity magazine and remarkable rate of fire made it a formidable weapon on the battlefield. Its success led to further advancements in submachine gun technology, influencing future designs such as the German MP40 and Soviet PPSh-41.

However, it was with his revolutionary anti-tank rifle that Lahti truly made his mark on weapon design. The Lahti L-39 anti-tank rifle was developed during World War II to counter armored vehicles used by opposing forces. Weighing over 100 pounds and firing a massive 20mm armor-piercing round, this rifle was capable of penetrating even heavily armored tanks at considerable distances. Its introduction transformed infantry tactics by providing soldiers with an effective means to neutralize enemy tanks from a safe distance.

Beyond their immediate impact on warfare, Lahti's designs laid the foundation for future developments in weapon technology. The principles he established continue to influence contemporary firearms design in terms of ergonomics, performance characteristics, and materials used. His emphasis on reliability and versatility has become standard practice when designing modern weaponry.

Moreover, Lahti's dedication to creating weapons that were effective yet practical for soldiers directly impacted military strategies during conflicts. By providing troops with powerful and reliable firearms, he empowered infantry units to engage enemy forces more effectively, ultimately influencing the outcome of battles.

Lahti's influence extended beyond his native Finland, as his designs found their way into the hands of various armed forces worldwide. The Suomi submachine gun became particularly popular during World War II and was adopted by several countries, including Sweden and Brazil. The Lahti L-39 anti-tank rifle also saw international use, proving its effectiveness in conflicts such as the Winter War between Finland and the Soviet Union.

Aimo Lahti's impact on weapon design is undeniable. His groundbreaking inventions not only revolutionized firearm technology but also influenced military strategies during armed conflicts. Lahti's legacy lives on through his contributions to modern submachine gun design and the development of the fearsome anti-tank rifle. His commitment to creating reliable and effective weapons continues to shape warfare today, ensuring that his name remains synonymous with innovation in weaponry.

***

Aimo Johannes Lahti, a Finnish soldier, renowned firearms designer, and national hero, left an indelible mark on Finland's military history. Throughout his illustrious career, Lahti received several prestigious honors and awards in recognition of his exceptional contributions to the field of weaponry. These accolades not only celebrated his remarkable skills as a designer but also acknowledged his unwavering dedication to his country.

One of the most notable honors bestowed upon Aimo Lahti was the Order of the Cross of Liberty. Established in 1918 during the Finnish Civil War, this esteemed award recognizes individuals for their extraordinary acts of bravery and service during times of war or national crisis. Lahti received this honor for his invaluable contributions to Finland's defense forces during World War II.

His innovative designs played a crucial role in equipping Finnish soldiers with highly effective firearms that helped them repel Soviet invasions.

In addition to the Order of the Cross of Liberty, Aimo Lahti was also awarded the Mannerheim Cross. Named after Field Marshal Carl Gustaf Emil Mannerheim, who served as Finland's military leader during World War II, this honor is regarded as one of Finland's highest military decorations. It is awarded to individuals who have displayed exceptional bravery and valor in combat situations.

By receiving this prestigious award, Lahti solidified his position as one of Finland's greatest war heroes.

Furthermore, Aimo Lahti was honored with the Pro Finlandia medal for outstanding artistic achievements that significantly contributed to promoting Finnish culture and identity abroad. While primarily known for his remarkable firearms designs rather than artistic pursuits, this recognition highlights not only his technical expertise but also how he elevated Finnish craftsmanship on a global scale.

Lahti's contributions were not limited to just weaponry design; he also made significant advancements in firearm ergonomics and safety features. In acknowledgment of these achievements, he received the Kultaristi (Gold Cross) medal, which is awarded to individuals who have made exceptional contributions to improving the functionality and usability of firearms. This recognition further exemplifies Lahti's commitment to ensuring the safety and effectiveness of weapons used by Finnish soldiers.

Finally, Aimo Lahti received the Finnish State Prize for Technology, a prestigious award that celebrates individuals who have made significant technological advancements with substantial societal impact. Lahti's innovative designs revolutionized firearm manufacturing processes and set new standards in terms of reliability, durability, and performance. By receiving this accolade, he was acknowledged not only as a pioneer in weaponry design but also as a technologist whose achievements transcended his field.

Aimo Lahti's honors and awards stand as testaments to his exceptional skills, bravery, and unwavering dedication to Finland. They recognize not only his contributions during times of conflict but also his invaluable impact on firearm design and technology. Through these accolades, Finland officially acknowledged Lahti as a national hero whose legacy continues to inspire generations of military personnel and firearms enthusiasts alike.

***

Aimo Johannes Lahti, a legendary Finnish soldier and firearm designer, had an extraordinary life that spanned both World Wars. Born on April 28, 1896, in Viiala, Finland, Lahti's journey from a young soldier to a national hero is one of immense valor and innovation. This subtopic aims to delve into the remarkable events that shaped his life before and after World War II.

Before the outbreak of World War II, Aimo Lahti had already established himself as an accomplished marksman. He began his military career in 1915 when he joined the Finnish Army during the Finnish Civil War. His exceptional shooting skills caught the attention of his superiors, leading him to become a sniper instructor during this period.

After the end of the war, Lahti focused on further honing his firearms knowledge. He traveled extensively across Europe to study various weapons systems and military tactics employed by different nations. This experience exposed him to cutting-edge technology and weaponry that would later influence his designs.

As tensions escalated before World War II, Finland sought to strengthen its defense capabilities. In 1931, Aimo Lahti was appointed as Chief Designer at Tikkakoski Arms Factory - Finland's primary firearms manufacturer. It was during this time that he created his most iconic invention: the Suomi KP/-31 submachine gun.

The Suomi KP/-31 became widely recognized for its reliability and exceptional accuracy even in challenging weather conditions – a crucial advantage for Finnish soldiers during their battles against Soviet forces during the Winter War (1939-1940). The weapon's effectiveness played a significant role in bolstering Finnish morale while capturing international attention due to its innovative design features.

Lahti's reputation as an ingenious firearm designer continued to grow after World War II. He went on to create other notable weapons such as the Lahti L-35 pistol, which gained recognition for its exceptional build quality and reliability. The pistol was widely adopted by the Finnish Army and saw use during the Continuation War (1941-1944).

Beyond his contributions to weapon design, Aimo Lahti also played a crucial role in post-war Finland. He actively participated in the demobilization process, helping veterans reintegrate into civilian life. Lahti's commitment to his fellow soldiers extended beyond his military duties, demonstrating his deep sense of duty and compassion.

Aimo Lahti's legacy extends far beyond his military achievements. His innovations in firearm design revolutionized the industry and left an indelible mark on Finnish military history. Today, he is remembered as a national hero, not only for his contributions during times of conflict but also for his dedication to improving the lives of those who served alongside him.

Aimo Lahti's life before and after World War II was marked by exceptional skills as a marksman and unparalleled innovation as a firearm designer. From humble beginnings in the Finnish Army to becoming a legend in Finnish military history, Lahti's impact will forever be etched in the annals of bravery and ingenuity.

***

Aimo Lahti, a legendary figure in Finnish history, is renowned for his exceptional contributions to the nation's military and engineering prowess. Born on April 28, 1896, in Viiala, Finland, Lahti's journey from a soldier to a legend is etched in the annals of Finnish history. His innovative mind and patriotic spirit have left an indelible mark on both Finland's military achievements and its cultural heritage.

Lahti's most notable contribution to Finnish history lies in his groundbreaking design of the iconic Suomi submachine gun. During World War II, Finland faced immense challenges as it fought against Soviet forces. Recognizing the need for an efficient weapon capable of withstanding harsh winter conditions, Lahti designed the Suomi submachine gun. Its reliability and effectiveness became legendary among soldiers on the front lines.

The impact of Lahti's invention extended far beyond its military application. The Suomi submachine gun symbolized Finland's resilience and determination to defend its sovereignty against overwhelming odds. It became an emblematic representation of national pride during wartime struggles. Today, this remarkable firearm stands as a testament to Aimo Lahti's ingenuity and continues to be revered as an important artifact in Finnish military history.

Beyond his contributions as an engineer, Aimo Lahti embodied the spirit of Finnish patriotism throughout his life. As a soldier himself during World War I and later serving as Lieutenant Colonel in the Finnish Army during World War II, he demonstrated unwavering dedication to defending his homeland against external threats. His commitment to duty inspired countless others within Finland's armed forces.

Lahti also played a significant role in shaping Finland's cultural heritage through his involvement with various artistic endeavors. An accomplished sculptor, he created several statues that celebrate key moments in Finland's history and honor those who sacrificed their lives for their country. These sculptures serve as visual reminders of Finland's rich past and the courage displayed by its people.

In addition to his artistic contributions, Lahti's legacy is preserved through the Aimo Lahti Foundation. Established in his honor, this foundation supports young Finnish engineers and innovators, ensuring that Lahti's spirit of innovation lives on. By providing scholarships and grants, the foundation continues to inspire future generations to follow in Aimo Lahti's footsteps.

Aimo Lahti's life and achievements have become an integral part of Finnish history and culture. His inventions revolutionized Finnish military capabilities while symbolizing resilience and determination. Through his artistic endeavors and the establishment of a foundation in his name, he has left an enduring legacy that resonates with both current and future generations. As Finland remembers Aimo Lahti, it celebrates not only a soldier turned legend but also a national hero who embodied the essence of Finnish patriotism.

***

The Suomi submachine gun is a legendary weapon that played a significant role in shaping military history. Designed by Finnish soldier Aimo Lahti, this remarkable firearm not only propelled him to fame but also became an iconic symbol of Finnish resistance during World War II. The success of the Suomi submachine gun can be attributed to several key factors, including its innovative design, exceptional performance, and strategic implementation.

One of the secrets behind the Suomi's triumph lies in its groundbreaking design. Aimo Lahti envisioned a weapon that could withstand extreme conditions and deliver unparalleled firepower. The submachine gun featured a sturdy construction with high-quality materials, making it reliable even in harsh environments. Its compact size allowed for easy handling and maneuverability, enabling soldiers to operate effectively in close-quarter combat situations.

Moreover, Lahti incorporated ingenious engineering techniques into the Suomi's design that significantly enhanced its performance. The submachine gun was equipped with a progressive trigger mechanism, allowing for both semi-automatic and fully automatic firing modes. This feature provided soldiers with flexibility during engagements, enabling them to adapt their tactics based on the circumstances they faced on the battlefield.

Another crucial aspect contributing to the success of the Suomi was its exceptional range and accuracy. Unlike many other submachine guns at that time, which were primarily designed for short-range engagements, Lahti aimed to create a weapon capable of engaging targets at longer distances effectively. By incorporating an extended barrel length and using 9mm Parabellum ammunition rather than smaller pistol cartridges typically used in submachine guns, he achieved remarkable accuracy and increased effective range.

Furthermore, Finland's strategic implementation of the Suomi played a pivotal role in solidifying its reputation as an outstanding firearm. During World War II, when Finland faced

Soviet aggression during what is known as the Winter War (1939-1940), soldiers armed with Suomi submachine guns demonstrated their superiority in combat. Finnish troops utilized guerrilla tactics and utilized the Suomi's exceptional firepower and range advantage to great effect, inflicting heavy casualties on Soviet forces and gaining a tactical edge.

The success of the Suomi submachine gun went beyond its technical capabilities; it also became a symbol of Finnish resilience and determination. The weapon's reputation spread worldwide, capturing the attention of military enthusiasts and professionals alike. Its effectiveness in combat, combined with its distinctive appearance, made the Suomi an iconic representation of Finnish resistance against overwhelming odds.

The Suomi submachine gun owes its success to its innovative design, exceptional performance, and strategic implementation during World War II. Aimo Lahti's vision for a reliable weapon capable of delivering superior firepower resulted in a firearm that surpassed expectations. With its remarkable accuracy, extended range, and adaptability on the battlefield, the Suomi became an instrument of both military prowess and national pride for Finland.

The legacy of Aimo Lahti lives on through his creation as soldiers continue to revere the Suomi submachine gun as one of history's most legendary firearms.

Aimo Lahti, a Finnish soldier turned legendary weapon designer, left an indelible mark on the world through his exceptional craftsmanship and dedication to innovation. His contributions to the field of firearms not only revolutionized military technology but also reflected his unwavering commitment to ensuring the safety and effectiveness of soldiers in battle. The enduring influence of Lahti's life and work can be seen in his remarkable designs, which continue to be revered by military enthusiasts, collectors, and historians worldwide.

Lahti's most iconic creation, the Suomi submachine gun, stands as a testament to his ingenuity and meticulous attention to detail. Its reliability, firepower, and ease of use set new standards for submachine guns during its time. Even today, more than eight decades after its introduction, the Suomi remains highly regarded for its exceptional performance on the battlefield. Its influence can be observed in subsequent firearms designs that drew inspiration from Lahti's innovative approach.

Beyond his contributions to weaponry design, Lahti's legacy extends into other areas as well. His passion for engineering excellence prompted him to explore various fields beyond firearms. From designing bicycles to developing machinery for agricultural purposes, he demonstrated versatility and an unwavering commitment to craftsmanship across industries.

Furthermore, Aimo Lahti's dedication to improving soldiers' lives went beyond designing weapons; he also played a crucial role in developing protective gear for troops during World War II. His gas mask design became widely adopted by Finnish soldiers due to its superior filtration capabilities and comfortable fit. This innovation not only saved numerous lives but also showcased Lahti's deep empathy for those who served their country.

Lahti's impact goes far beyond his tangible creations; he was an inspirational figure whose work ethic continues to inspire generations of engineers today. His relentless pursuit of excellence serves as a reminder that true success is achieved through hard work, perseverance, and unwavering dedication to one's craft. Lahti's legacy serves as a testament to the power of passion and determination in shaping the world.

Moreover, Aimo Lahti's influence extends beyond his design contributions, as his work has become an integral part of Finland's rich history. His name is synonymous with Finnish engineering excellence, and his creations are proudly displayed in museums across the country. Lahti's designs have become national treasures that symbolize Finland's commitment to innovation and technological advancement.

Aimo Lahti's life and work have left an enduring impact on both military technology and Finnish history. His remarkable designs continue to be celebrated for their innovation, reliability, and effectiveness. Moreover, his dedication to craftsmanship and commitment to improving soldiers' lives serve as a source of inspiration for engineers worldwide. Aimo Lahti will forever be remembered as a legend whose contributions have shaped the world we live in today.

Yisrael Galili

1911-1986

In the tumultuous years following Israel's establishment as an independent nation, one man emerged as a key figure in shaping the country's defense industry. Yisrael Galili, a visionary leader and innovator, played a crucial role in transforming Israel's military capabilities and establishing it as a global powerhouse in defense technology. From his humble beginnings to his remarkable achievements, Galili's contributions have left an indelible mark on the Israeli defense landscape.

Born in Haifa, Palestine, in 1923, Yisrael Galili grew up amidst the backdrop of political upheaval and uncertainty. Witnessing firsthand the struggles faced by Jewish immigrants seeking refuge from persecution around the world, he developed an unwavering determination to protect and defend his people. This early exposure to adversity fueled his passion for ensuring Israel's security and set him on a path that would redefine its defense industry.

Galili began his career as an engineer at Rafael Advanced Defense Systems (then known as HEMED), one of Israel's leading defense research institutions. It was here that he honed his skills and gained invaluable experience that would shape his future endeavors. Galili quickly distinguished himself with his innovative thinking and problem-solving abilities, earning him recognition within the organization.

However, it was during the Six-Day War in 1967 that Galili truly showcased his exceptional leadership qualities. Serving as Director-General of the Ministry of Defense during this critical period for Israel's survival, he played a pivotal role in ensuring military preparedness by overseeing strategic planning efforts. His expertise enabled rapid mobilization of troops and resources while coordinating intelligence operations crucial to securing victory against multiple enemy forces.

Galili recognized that technological advancements were vital for maintaining Israel's military edge amidst regional conflicts. As such, he spearheaded initiatives aimed at developing cutting-edge weaponry systems tailored specifically to address Israel's unique security challenges. Under his leadership, Israel's defense industry flourished, resulting in the creation of groundbreaking technologies such as the Merkava battle tank and the Arrow missile defense system.

Beyond his contributions to Israel's defense capabilities, Galili also played a crucial role in establishing international collaborations that bolstered Israel's position on the global stage. His strategic partnerships with several countries facilitated knowledge exchange and joint research ventures that further enhanced Israel's technological prowess.

Yisrael Galili's legacy extends far beyond his professional achievements. Throughout his life, he remained committed to fostering a culture of innovation and excellence within the Israeli defense industry. His unwavering dedication to protecting the people of Israel ensured their safety and established a foundation for future advancements in defense technology.

As we delve deeper into Yisrael Galili's remarkable journey through this text, we will explore the pivotal moments that shaped him as a pioneer in the Israeli defense industry. From his role in transforming Rafael into a world-class institution to his contributions during times of conflict, Galili's story serves as an inspiration for generations to come.

***

Yisrael Galili, a pioneer in the Israeli defense industry, played a crucial role in the establishment and development of Israel's defense capabilities. Born on September 10, 1923, in Tel Aviv, Galili grew up during a tumultuous time for the Jewish people in Palestine. His early years were marked by his deep commitment to safeguarding the Jewish homeland.

Galili's passion for defending his people began at a young age. As a teenager, he joined the Haganah, a paramilitary organization that aimed to protect Jewish settlements from Arab attacks. This experience exposed him to the realities of conflict and instilled in him an unwavering determination to ensure Israel's security.

After completing high school, Galili decided to pursue his higher education at Technion – Israel Institute of Technology. He studied mechanical engineering with a specialization in military technology. His academic pursuits provided him with valuable knowledge and skills that would prove essential in shaping his future contributions to the defense industry.

Upon graduating from Technion in 1946, Galili dedicated himself fully to serving his country. He joined the Palmach – an elite unit within the Haganah – where he honed his military skills and leadership abilities. During this time, he actively participated in several operations against Arab forces seeking to undermine Jewish settlements.

Galili's exceptional performance did not go unnoticed by David Ben-Gurion, Israel's first prime minister. Recognizing Galili's talent and expertise, Ben-Gurion appointed him as one of three officers responsible for organizing and developing Israel's military industries infrastructure.

In this new role, Galili faced numerous challenges as he worked tirelessly to establish an independent defense industry capable of meeting Israel's unique security needs. He played a vital part in setting up factories for ammunition production and small arms manufacturing within a short period.

Moreover, Galili focused on creating a strong research and development sector within the defense industry. He facilitated collaborations between Israeli scientists, engineers, and military personnel to develop cutting-edge technologies essential for Israel's defense. His efforts laid the foundation for Israel's future advancements in weaponry, surveillance systems, and communication networks.

Galili's dedication to innovation and his commitment to ensuring Israel's security led to significant breakthroughs in military technology. His visionary leadership played a crucial role in transforming Israel from a nation heavily reliant on foreign arms imports into a self-sufficient powerhouse in defense manufacturing.

The early years of Galili's career marked the birth of the Israeli defense industry. His unwavering determination, expertise in military technology, and visionary leadership were instrumental in shaping Israel's ability to defend itself effectively. Galili's contributions continue to resonate today as Israel remains at the forefront of technological advancements in the global defense industry.

Yisrael Galili is widely recognized as one of the pioneers in the Israeli defense industry, playing a significant role in advancing military technology in the country. Through his innovative thinking and relentless pursuit of excellence, Galili made groundbreaking contributions that revolutionized Israel's defense capabilities. This subtopic will delve into some of his notable achievements and their impact on military technology advancements in Israel.

One of Galili's most remarkable contributions was the development of the Uzi submachine gun. In collaboration with Major Uziel Gal, he designed this compact weapon that quickly became an iconic symbol of Israeli military prowess. The Uzi's lightweight and reliable design made it highly effective for close-quarters combat, providing Israeli soldiers with a significant advantage on the battlefield. Its success led to its adoption by various armed forces worldwide, solidifying Israel's reputation as a leader in firearms technology.

Galili also played a crucial role in developing innovative armored vehicles for the Israeli Defense Forces (IDF). He recognized that traditional tanks were vulnerable to anti-tank missiles and sought to create a new generation of armored vehicles that could withstand these threats. His efforts resulted in the creation of Merkava, a revolutionary battle tank equipped with advanced armor protection systems and state-of-the-art technologies.

The Merkava became an integral part of Israel's military arsenal, offering enhanced survivability for its crew and setting new standards for tank design worldwide.

In addition to firearms and armored vehicles, Galili contributed significantly to Israel's air defense capabilities. He led efforts to develop surface-to-air missile systems that could effectively neutralize aerial threats posed by enemy aircraft or missiles. Under his guidance, Israel successfully developed advanced missile defense systems such as Iron Dome, Arrow II/III, and David's Sling. These systems have proven instrumental in safeguarding Israeli airspace against hostile attacks, ensuring the country's security and deterring potential aggressors.

Furthermore, Galili's vision extended beyond individual weapon systems. He recognized the importance of integrating various technologies to enhance overall military effectiveness. Galili spearheaded projects that focused on developing advanced command and control systems, intelligence gathering capabilities, and cyber defense measures. By prioritizing these areas, he paved the way for Israel to become a global leader in network-centric warfare and cybersecurity.

Galili's contributions to military technology advancements in Israel have had a profound impact on the country's defense capabilities. His innovative designs and strategic thinking not only provided Israeli forces with cutting-edge weaponry but also bolstered their overall operational efficiency. Through his pioneering efforts, Galili set new standards for excellence within the Israeli defense industry, inspiring future generations of engineers and scientists to continue pushing boundaries in military technology.

Yisrael Galili's contributions to military technology advancements in Israel are nothing short of remarkable. From designing iconic firearms like the Uzi to developing groundbreaking armored vehicles and missile defense systems, his ingenuity has shaped Israel's defense industry. His legacy serves as an enduring testament to the importance of innovation and dedication in ensuring a nation's security.

***

Yisrael Galili, a renowned pioneer in the Israeli defense industry, has made significant contributions to the development of advanced weaponry that have played a crucial role in safeguarding the nation's security. Throughout his career, Galili spearheaded several groundbreaking projects and achieved remarkable breakthroughs that revolutionized Israel's defense capabilities. This article explores some of his key achievements in weapons development.

One of Galili's most notable breakthroughs was the invention of the Uzi submachine gun, which became an iconic symbol of Israeli military strength. The Uzi was designed to be compact, lightweight, and reliable, making it highly effective for close-quarter combat situations. Its innovative design incorporated a telescoping bolt system that reduced recoil and provided excellent accuracy even during rapid-fire sequences. The Uzi quickly gained popularity both domestically and internationally due to its exceptional performance and ease of use.

Galili also played a pivotal role in developing the Negev light machine gun, which addressed the evolving needs of modern warfare. This weapon system incorporated advanced materials and ergonomics to provide increased firepower while maintaining maneuverability on the battlefield. The Negev utilized an open-bolt mechanism combined with quick-change barrels to enhance sustained fire capabilities without compromising reliability or accuracy. It became an essential asset for Israeli infantry units operating in various combat scenarios.

Another groundbreaking achievement by Galili was his involvement in the creation of Israel's first domestically produced assault rifle – the Galil. This weapon system combined elements from various renowned rifles worldwide while integrating innovative features tailored for Israeli military requirements. The Galil boasted exceptional reliability even under harsh conditions and featured a robust gas piston operating system that reduced fouling and improved overall performance.

Its versatility allowed soldiers to adapt quickly between different combat roles with ease.

Galili also contributed significantly to missile technology advancements in Israel by overseeing critical projects such as the Gabriel anti-ship missile program. The Gabriel missile system provided the Israeli Navy with a formidable weapon capable of engaging hostile naval forces at extended ranges. Its advanced radar guidance and sea-skimming capabilities made it extremely difficult for enemy vessels to detect and counter. The success of the Gabriel missile program solidified Israel's position as a global leader in missile technology.

Furthermore, Galili played a vital role in developing innovative armored vehicle technologies that enhanced Israel's ground capabilities. He spearheaded projects such as the Merkava tank, which introduced several groundbreaking features including modular armor, rear-mounted engine configuration, and advanced crew protection systems. The Merkava revolutionized armored warfare by prioritizing crew survivability and providing unmatched mobility and firepower.

Yisrael Galili's key achievements in weapons development have had a profound impact on Israel's defense industry. His inventions not only provided Israeli forces with cutting-edge weaponry but also established Israel as a leading innovator in defense technologies globally. Galili's contributions continue to shape the nation's security landscape, ensuring its preparedness against evolving threats while safeguarding its sovereignty.

***

Yisrael Galili was a visionary and pioneer in the Israeli defense industry, playing a crucial role in strengthening Israel's national security. Born on October 10, 1923, in Tel Aviv, Galili dedicated his life to ensuring the safety and protection of his homeland. Through his relentless efforts and groundbreaking contributions, he significantly bolstered Israel's military capabilities and played a pivotal role in its defense strategy.

Galili began his career as a combat soldier during Israel's War of Independence in 1948. This firsthand experience on the frontlines gave him valuable insights into the challenges faced by Israeli forces. Recognizing the need for advanced weaponry and technology, Galili focused his efforts on research and development to enhance Israel's defense capabilities.

One of Galili's most significant contributions was his instrumental role in establishing the Israeli Military Industries (IMI) corporation. Founded in 1933 as a small ammunition factory, IMI evolved under Galili's leadership to become one of the world's leading developers and manufacturers of cutting-edge defense systems. The corporation played an essential role in equipping Israeli forces with state-of-the-art weapons, ammunition, and military technology.

Under Galili's guidance as Director-General from 1962 to 1974, IMI developed several iconic weapons that became vital assets for Israel's national security. One such innovation was the Uzi submachine gun. Compact yet powerful, it quickly gained international recognition for its reliability and effectiveness on the battlefield. The Uzi became synonymous with Israeli military prowess worldwide.

Moreover, Galili spearheaded research into missile technology at IMI during a time when this field was still relatively nascent. His efforts laid the foundation for Israel to develop advanced missile systems that remain integral components of its national security strategy today. By investing heavily in research and development, Galili helped Israel become a regional leader in missile defense.

Galili's contributions extended beyond the development of weapons and technology. He recognized the importance of forging international partnerships to further enhance Israel's national security. By fostering collaborations with other countries, Galili facilitated the exchange of knowledge, resources, and expertise. These alliances not only strengthened Israel's military capabilities but also fostered diplomatic ties that continue to benefit the nation to this day.

Furthermore, Galili played a key role in shaping Israel's defense policy by advocating for innovative strategies and approaches. He emphasized the need for a multidimensional approach to security that encompassed not only military strength but also intelligence gathering and cybersecurity measures. His forward-thinking vision led to significant advancements in these areas, ensuring Israel remained at the forefront of emerging security challenges.

Yisrael Galili was an exceptional leader whose contributions revolutionized Israel's defense industry and national security landscape. Through his dedication, innovation, and strategic vision, he strengthened Israel's military capabilities while forging crucial international partnerships. Galili's legacy continues to resonate today as his groundbreaking work remains foundational to Israel's ongoing commitment to protecting its citizens and maintaining regional stability.

***

Yisrael Galili, a pioneer in the Israeli defense industry, played a pivotal role in establishing collaborations and partnerships with various countries, thereby significantly influencing international defense alliances. Galili's visionary leadership and expertise in military technology propelled Israel's defense sector to new heights, fostering global cooperation and exchange of knowledge.

One of Galili's most notable contributions was his instrumental role in forging a strategic alliance between Israel and the United States. Recognizing the potential for mutual benefit, he initiated discussions that led to the establishment of joint research and development projects between Israeli companies and their American counterparts. This collaboration not only enhanced Israel's military capabilities but also bolstered its standing as a key player in global defense affairs.

The partnership provided Israel access to advanced technologies while enabling American companies to tap into Israel's innovation-driven defense industry.

Galili also sought partnerships beyond traditional allies, realizing that cooperation with emerging powers could yield substantial benefits for both sides. His efforts resulted in fruitful collaborations with countries such as India, South Korea, and Singapore. Through these alliances, Israel provided cutting-edge military technology while gaining valuable insights into regional security challenges faced by its partners. These partnerships led to joint exercises, information sharing initiatives, and collaborative research projects that strengthened the collective security of all involved nations.

Furthermore, Galili actively promoted multinational collaborations within Europe through his engagement with NATO member states. Recognizing the need for collective security measures against common threats, he advocated for increased cooperation among European countries on defense matters. As a result of his efforts, several joint research programs were established between Israeli companies and European institutions specializing in defense technology development. This exchange of expertise paved the way for closer ties between Israel and European nations while fostering technological advancements across borders.

Galili's influence extended beyond traditional state-to-state collaborations; he also fostered connections with private sector entities involved in defense industries worldwide. Recognizing the importance of harnessing innovation from various sources, he facilitated partnerships between Israeli defense companies and multinational corporations. These collaborations not only enhanced the global reach of Israel's defense industry but also enabled knowledge transfer and technological advancements through joint research and development projects.

Yisrael Galili's visionary leadership and expertise in the Israeli defense industry had a profound impact on international defense alliances. Through his efforts, Israel forged strategic partnerships with countries such as the United States, India, South Korea, Singapore, and European nations. Additionally, Galili promoted collaboration between Israeli defense firms and multinational corporations worldwide. These collaborations not only enhanced Israel's military capabilities but also fostered global cooperation in defense technology development.

Galili's legacy as a pioneer in the Israeli Defense Industry continues to shape international defense alliances today.

***

Yisrael Galili, a pioneer in the Israeli Defense Industry, revolutionized modern warfare tactics through his innovative contributions. Galili's groundbreaking ideas and inventions have had a profound impact on military strategies and continue to shape the way armed forces operate today.

One of Galili's most significant contributions was the development of the Uzi submachine gun. Introduced in the 1950s, the Uzi quickly became an iconic weapon known for its compact size, simplicity, and reliability. Its lightweight design made it highly portable and easy to handle in close-quarter combat situations. The Uzi's high rate of fire gave soldiers a superior advantage on the battlefield, allowing them to engage multiple targets quickly and effectively.

This innovation transformed infantry tactics by providing soldiers with a versatile weapon that could be used for both offensive and defensive purposes.

Galili also played a key role in advancing armored vehicle technology through his development of the Merkava tank. Designed with crew survivability as its top priority, this battle tank featured innovative features such as rear-engine placement, modular armor systems, and an advanced suspension system.

These advancements greatly enhanced crew protection against anti-tank weaponry while maintaining excellent maneuverability in various terrains.

The Merkava tank's unique design also incorporated a rear door that allowed for rapid deployment of troops during combat operations. This innovation enabled infantry units to swiftly enter or exit the battlefield while under heavy fire, reducing their vulnerability during critical moments. By combining firepower with troop mobility, Galili transformed armored warfare tactics by creating an integrated system that maximized both offensive capabilities and soldier safety.

In addition to his contributions to small arms weaponry and armored vehicles, Galili made significant advancements in electronic warfare systems. He played an instrumental role in developing electronic countermeasures (ECM) technology that disrupts enemy communication systems and radar capabilities effectively. These ECM systems have become indispensable tools for modern militaries, enabling forces to gain a significant tactical advantage by neutralizing the enemy's ability to coordinate their actions effectively.

Furthermore, Galili's innovations in surveillance and reconnaissance technologies have greatly enhanced intelligence gathering capabilities on the battlefield. His contributions include the development of unmanned aerial vehicles (UAVs) that provide real-time situational awareness, enabling commanders to make informed decisions based on accurate and up-to-date information. These UAVs have become invaluable assets in modern warfare, supporting various operations such as target acquisition, monitoring enemy movements, and conducting aerial reconnaissance.

Yisrael Galili's pioneering efforts in the Israeli Defense Industry have had a profound impact on modern warfare tactics. Through his innovative contributions to small arms weaponry, armored vehicles, electronic warfare systems, and surveillance technologies, Galili revolutionized military strategies. His inventions continue to shape the way armed forces operate today by providing soldiers with advanced tools that enhance their effectiveness on the battlefield while prioritizing their safety.

***

Yisrael Galili, a true pioneer in the Israeli defense industry, has left an indelible mark on the nation's security landscape. His exceptional contributions have not only shaped Israel's defense capabilities but have also inspired and paved the way for future generations of defense experts.

Galili's unwavering commitment to innovation and excellence serves as a beacon for aspiring Israeli defense professionals. Throughout his illustrious career, he consistently pushed boundaries and challenged conventional wisdom, setting new standards in defense technology. His groundbreaking work has become the cornerstone of Israel's military prowess.

One of Galili's most significant contributions lies in his role as the driving force behind Israel's development of advanced missile systems. Recognizing the need for robust missile defenses to protect Israel from regional threats, he initiated and led several groundbreaking projects. His visionary leadership led to the creation of state-of-the-art missile interceptors that have proven their effectiveness time and again. Galili's legacy in this field continues to inspire young engineers and scientists to develop innovative solutions to safeguard their nation.

Furthermore, Galili played a pivotal role in fostering collaboration between academia, industry, and government agencies. He understood that true progress could only be achieved through synergistic partnerships. By facilitating communication and knowledge-sharing among these different entities, Galili helped create an ecosystem conducive to breakthrough research and development. His emphasis on interdisciplinary collaboration has become a guiding principle for future generations seeking to tackle complex defense challenges.

Galili was not just an accomplished engineer; he was also a mentor who nurtured talent within the Israeli defense industry. He believed that investing in human capital was crucial for maintaining technological superiority. Through his guidance and mentorship programs, he imparted invaluable knowledge and skills to aspiring professionals, ensuring continuity in expertise within the sector.

Moreover, Galili's unwavering commitment to ethical conduct serves as an enduring example for future generations. In an industry where integrity is of paramount importance, Galili always upheld the highest standards of professionalism and ethics. His dedication to transparency and accountability has become a guiding light for aspiring defense experts, emphasizing the importance of upholding moral values while safeguarding national security.

Today, Galili's legacy continues to inspire the next generation of Israeli defense experts. As they embark on their own journeys in the defense industry, they carry with them his innovative spirit, commitment to collaboration, investment in human capital, and unwavering ethical conduct. Galili's influence reverberates through research laboratories and engineering departments across Israel as young minds strive to build upon his achievements.

Yisrael Galili's pioneering contributions have left an indelible mark on Israel's defense industry. His visionary leadership and technological breakthroughs have not only enhanced Israel's security but also inspired future generations of defense experts. As these young professionals strive to follow in his footsteps, they are driven by Galili's legacy – a legacy that embodies innovation, collaboration, mentorship, and ethical conduct – ensuring that Israel remains at the forefront of global defense capabilities for years to come.

***

Yisrael Galili, a visionary and pioneer in the Israeli defense industry, encountered numerous challenges throughout his illustrious career. From developing innovative military technologies to establishing successful defense companies, Galili's journey was marked by obstacles that shaped his approach to business and leadership. This subtopic explores some of the key challenges faced by Galili and the invaluable lessons he learned along the way.

One of the primary challenges Galili encountered was navigating a highly competitive global market. In an industry dominated by established military powers, breaking through as a newcomer required exceptional determination and strategic thinking. Recognizing this hurdle, Galili focused on leveraging Israel's reputation for technological prowess to create cutting-edge solutions that would capture international attention. He understood that differentiation would be crucial in gaining a competitive edge.

Another significant challenge faced by Galili was establishing credibility as a defense contractor in an era when Israel was still emerging as a nation-state. Building trust with potential clients meant overcoming skepticism about Israel's capabilities while simultaneously proving their worthiness through tangible results. This challenge forced Galili to prioritize quality assurance and deliver products that consistently exceeded expectations—a lesson he carried throughout his career.

Galili also grappled with financial constraints during his early years in business. Funding research and development initiatives required significant capital investment, which was not always readily available to him. To overcome this challenge, he sought partnerships with both domestic and international investors who shared his vision for innovation in defense technologies. These collaborations not only provided much-needed financial support but also helped establish networks that proved beneficial for future ventures.

Furthermore, navigating complex bureaucratic processes posed an ongoing challenge for Yisrael Galili within the defense industry. Dealing with government regulations, export controls, and security protocols required patience and persistence at every stage of development and production. By recognizing the importance of establishing strong relationships with government officials, Galili effectively minimized bureaucratic hurdles and ensured smooth operations. This experience taught him the value of building connections within the industry and maintaining open lines of communication with relevant authorities.

A crucial lesson learned by Galili was the significance of adaptability in an ever-evolving defense landscape. Technological advancements and changing geopolitical dynamics demanded constant innovation to stay ahead. Galili embraced a culture of continuous learning and encouraged his teams to remain at the forefront of cutting-edge research and development. This approach allowed him to anticipate emerging trends, develop new products, and respond swiftly to evolving customer needs.

Yisrael Galili's journey in the Israeli defense industry was not without its challenges. From fierce competition to financial constraints, bureaucratic hurdles to credibility issues, he faced numerous obstacles that tested his resilience as a pioneer. However, through determination, strategic thinking, quality assurance, partnerships, adaptability, and building strong relationships with relevant authorities, Galili overcame these challenges while learning invaluable lessons along the way.

His success serves as an inspiration for aspiring entrepreneurs in the defense industry who strive for excellence amidst adversity.

Throughout his illustrious career, Yisrael Galili emerged as a true pioneer and visionary leader in the Israeli defense industry.

His relentless dedication to strengthening Israel's military capabilities, coupled with his innovative thinking and strategic approach, played a pivotal role in shaping the nation's defense landscape. Galili's unwavering commitment to excellence, technological advancement, and national security has left an indelible mark on Israel's defense industry.

Galili's journey began with his service in the Palmach during the War of Independence. It was during this time that he developed a deep understanding of the importance of a strong defense infrastructure for Israel's survival. This experience laid the foundation for his lifelong mission to bolster Israel's military capabilities and ensure its security.

As a founding member of Rafael Advanced Defense Systems, Galili played a crucial role in establishing one of Israel's most prominent defense companies. Under his leadership as CEO, Rafael flourished into an internationally recognized organization renowned for its cutting-edge technologies. Galili fostered a culture of innovation within Rafael, encouraging collaboration among scientists and engineers to push boundaries and develop groundbreaking solutions.

One of Galili's most significant contributions was spearheading the development of Iron Dome, an advanced missile-defense system that has revolutionized modern warfare. Recognizing the threat posed by short-range rockets from neighboring territories, Galili championed this project from its inception to its successful deployment on the battlefield. Iron Dome has since saved countless lives by intercepting incoming projectiles before they can cause destruction or harm to innocent civilians.

Furthermore, Galili played a pivotal role in fostering collaborations between Israeli defense industries and international partners. He understood that by forging alliances with leading global players, Israel could enhance its technological capabilities while also contributing to global security efforts. Through joint ventures and technology transfers with countries around the world, Galili ensured that Israel remained at the forefront of defense innovation.

Galili's visionary leadership extended beyond the realm of technology and entrepreneurship. He understood that a strong defense industry required a well-trained and motivated workforce. As such, he championed educational initiatives, promoting science, technology, engineering, and mathematics (STEM) education to cultivate a new generation of skilled professionals in Israel's defense sector. Galili's emphasis on human capital development has had a lasting impact on the industry's growth and sustainability.

Yisrael Galili stands as an iconic figure in the history of the Israeli defense industry. His visionary leadership, innovative thinking, and unwavering commitment to national security have reshaped Israel's defense landscape. Through his pioneering efforts at Rafael Advanced Defense Systems and his instrumental role in the development of Iron Dome, Galili has left an enduring legacy that continues to safeguard Israel from emerging threats.

As Israel progresses into the future, it is essential to recognize Galili as a true pioneer whose contributions have solidified the nation's position as a global leader in defense technology and innovation.

Mikhail Kalashnikov

1919-2013

Mikhail Timofeyevich Kalashnikov, the renowned Russian inventor, was born on November 10, 1919, in Kurya, a small village located in the Altai Krai region of Western Siberia. His humble beginnings and challenging upbringing greatly influenced his perspective on life and ultimately shaped his destiny as a prolific firearms designer.

Growing up in a large peasant family, Kalashnikov experienced firsthand the hardships that came with living in rural Russia during the early 20th century. His father was a farmer who struggled to make ends meet, often forcing young Mikhail to work alongside him from an early age. Despite their financial struggles, his parents emphasized the importance of education and encouraged him to pursue knowledge whenever possible.

In 1930, tragedy struck when Kalashnikov's father died due to illness. This event marked a turning point in his life as it placed an even greater burden on him to support his family. As one of eighteen children, he felt a strong sense of responsibility towards his siblings and mother. To contribute to their livelihoods, he took on various odd jobs such as working at construction sites and joining logging expeditions.

However, fate would soon intervene when World War II erupted in 1941. Kalashnikov was drafted into the Red Army where he served as a tank mechanic despite having no prior military experience or training. It was during this time that he found himself inspired by Soviet soldiers' need for more reliable firearms against German weaponry.

During his recovery from injuries sustained in battle in 1942, Kalashnikov began sketching ideas for new firearm designs. Drawing upon his mechanical knowledge gained through repairing tanks and studying other weapons systems, he aimed to create a weapon that would be both durable and easy to manufacture—a weapon that could give Russian soldiers an advantage on the battlefield.

In 1947, after years of experimentation and refinement, Kalashnikov introduced his most significant creation—the Avtomat Kalashnikova model 1947, better known as the AK-47. This groundbreaking assault rifle would go on to become one of the most iconic firearms in history, renowned for its simplicity, reliability, and affordability.

Despite achieving worldwide recognition for his invention, Kalashnikov remained a modest and humble individual throughout his life. He devoted himself to advancing firearm technology and improving the lives of soldiers rather than seeking personal fame or fortune. He received numerous accolades for his contributions to weapon design but always maintained that his invention was intended solely for defensive purposes.

Mikhail Kalashnikov's early life experiences profoundly influenced his journey as an inventor. From a poverty-stricken upbringing in rural Russia to becoming a legendary figure in the world of firearms, he overcame adversity through determination and perseverance. His AK-47 revolutionized modern warfare and left an indelible mark on history—a legacy that will forever be associated with both innovation and controversy.

***

Mikhail Kalashnikov, a name synonymous with the AK-47, is one of the most renowned firearms designers in history. But what many may not know is that his journey from soldier to innovator played a crucial role in shaping his remarkable inventions. Kalashnikov's military experience provided him with firsthand knowledge of the challenges faced by soldiers on the battlefield, inspiring him to create weapons that were not only efficient and reliable but also easy to use and maintain.

Born in 1919 in Russia, Kalashnikov experienced the turmoil of World War II as a young soldier. He served as a tank commander during the war and witnessed first-hand the devastating effects of modern warfare. It was during this time that he realized the need for a more effective weapon for infantry soldiers – something that could provide them with an advantage over their adversaries.

Kalashnikov's military experience gave him valuable insights into the shortcomings of existing firearms, particularly in terms of reliability and simplicity. He observed how soldiers struggled with complex mechanisms and unreliable weapons when faced with harsh conditions on the battlefield. Determined to improve upon these flaws, he embarked on a journey to design a rifle that would revolutionize modern warfare.

The AK-47 (Avtomat Kalashnikova model 1947) was born out of this desire to create an efficient and reliable weapon suitable for all combat situations. Drawing from his own experiences as well as feedback from fellow soldiers, Kalashnikov meticulously designed every component of this iconic rifle. The result was a firearm that not only excelled in terms of performance but also surpassed expectations in terms of ease-of-use and maintenance.

One significant aspect influenced by his military background was the AK-47's ability to withstand extreme conditions without compromising functionality. Soldiers often operate under challenging environments such as deserts or freezing temperatures, where other firearms would fail. Kalashnikov's design ensured that the AK-47 could reliably function in any climate or terrain, ensuring soldiers could trust their weapon regardless of the circumstances.

Furthermore, Kalashnikov's military experience also played a pivotal role in shaping his commitment to simplicity. He understood that soldiers needed a weapon that could be easily operated and maintained, even with minimal training. The AK-47's design incorporated this philosophy by minimizing complex mechanisms and utilizing rugged materials to ensure durability. As a result, soldiers were able to quickly learn how to operate and maintain the rifle effectively, reducing downtime and increasing overall combat efficiency.

Mikhail Kalashnikov's military experience directly influenced his journey from soldier to innovator. His firsthand understanding of the challenges faced by infantry soldiers on the battlefield inspired him to create a firearm that not only addressed these challenges but exceeded expectations in terms of reliability and ease-of-use. The AK-47 stands as a testament to his dedication and commitment towards improving weaponry for the benefit of soldiers worldwide.

***

In the annals of firearms history, few names hold as much weight as Mikhail Kalashnikov. His brainchild, the AK-47 assault rifle, has become an iconic symbol of power, revolution, and warfare. The AK-47 revolutionized modern warfare and forever changed the course of history. This subtopic aims to delve into the birth of this legendary weapon and shed light on its enduring impact.

Born in 1919 in a remote village in Russia, Mikhail Kalashnikov displayed an early fascination with engineering. However, it was during his time in World War II that he truly honed his skills as a weapons designer. It was on the battlefields that he witnessed firsthand the limitations and flaws of existing rifles used by Soviet troops against their German adversaries.

Inspired by this experience, Kalashnikov set out to create a superior firearm that would offer reliability, simplicity, and unparalleled firepower.

It took several years of trial and error before Kalashnikov finally unveiled his masterpiece - the Avtomat Kalashnikova 1947 or AK-47 for short. Introduced in 1947 after extensive testing and refinement, this assault rifle instantly captured the attention of military strategists around the world. Its design incorporated several revolutionary features that set it apart from its contemporaries.

One key innovation was its use of intermediate cartridge ammunition - smaller than traditional rifle rounds but more powerful than pistol cartridges. This not only allowed for greater accuracy at longer ranges but also made it easier for soldiers to carry larger quantities of ammunition into battle.

Another remarkable feature was its gas-operated mechanism combined with a rotating bolt system. This design allowed for reliable operation even under harsh conditions such as extreme cold or excessive dirt buildup - two factors that had plagued earlier rifles.

However, what truly made the AK-47 exceptional was its simplicity. With just eight major components, it was easy to manufacture, maintain, and repair. This meant that even countries with limited industrial capabilities could produce their own versions of the AK-47, leading to its widespread adoption across the globe.

The AK-47's impact on warfare cannot be overstated. Its availability and effectiveness made it the weapon of choice for numerous liberation movements and insurgencies in the latter half of the 20th century. From Vietnam to Afghanistan, from Angola to Mozambique, the AK-47 became synonymous with guerrilla warfare and revolution.

Furthermore, its ubiquity has led to estimates that there are over 100 million AK-47 rifles in circulation today. Its affordability combined with its durability has made it a favorite among both state armies and non-state actors alike.

Mikhail Kalashnikov's genius gave birth to a weapon that forever changed the face of modern warfare - the AK-47 assault rifle. Through its revolutionary design and unparalleled reliability, this legendary firearm has become an enduring symbol of power and resistance. The AK-47 revolutionized warfare by offering a simple yet deadly solution for soldiers across the globe. Its impact continues to shape conflicts even today as it remains an indelible part of our collective consciousness.

***

Mikhail Kalashnikov, a Soviet weapons designer, forever changed the landscape of warfare with his revolutionary creation – the AK-47. This iconic assault rifle has not only become a symbol of power and conflict but also a testament to Kalashnikov's ingenuity and engineering prowess. Delving into the design process of the AK-47 provides us with an inside look at how this remarkable firearm came to be.

In the aftermath of World War II, Kalashnikov embarked on a quest to create a weapon that would outperform its predecessors in terms of reliability, simplicity, and firepower. His experience as a tank mechanic during the war allowed him to understand firsthand the importance of a reliable firearm in combat situations. With this vision in mind, he began designing what would later become one of the most successful firearms in history.

The key principle behind Kalashnikov's design philosophy was simplicity. He aimed to create a weapon that could be produced quickly and efficiently using minimal resources while still being effective on the battlefield. The AK-47 achieved this by utilizing stamped metal parts instead of intricate machining processes, reducing both cost and complexity.

One crucial aspect that made the AK-47 stand out was its remarkable reliability even under adverse conditions. The innovative gas-operated mechanism employed by Kalashnikov allowed for self-cleaning during firing, preventing malfunctions caused by dirt or debris buildup. This made it particularly suitable for soldiers operating in harsh environments where maintenance might be challenging.

Another notable feature was its exceptional firepower combined with ease of use. The intermediate cartridge used in the AK-47 struck an ideal balance between range and stopping power while minimizing recoil for accurate rapid-fire shooting. Moreover, its ergonomic design enabled soldiers to handle it effectively even without extensive training.

Throughout its development, Kalashnikov tirelessly refined his creation, continuously seeking feedback from soldiers and incorporating their suggestions. This iterative process ensured that the AK-47 was not only a product of his genius but also a result of collective experience and practicality.

The AK-47's impact on military history is undeniable. Its production began in 1947, and by the late 1950s, it had become the standard issue firearm for Soviet forces. Its effectiveness in combat led to its widespread adoption across many nations, fueling countless conflicts around the globe.

However, despite its association with violence and warfare, Kalashnikov always maintained that he created the AK-47 to protect his homeland rather than to cause harm. In later years, he expressed deep regret over the weapon's misuse and dedicated himself to promoting peace through educational initiatives.

Mikhail Kalashnikov's ingenious invention reshaped modern warfare forever. The simplicity, reliability, and firepower of the AK-47 made it an unparalleled success in combat situations. Beyond its functionality, this iconic firearm stands as a testament to human ingenuity and innovation – a remarkable accomplishment by one man that forever transformed the course of history.

***

The AK-47, also known as the Avtomat Kalashnikova model 1947, is widely regarded as one of the most influential weapons in modern history. Designed by Mikhail Kalashnikov, a Soviet weapons engineer, in the aftermath of World War II, this assault rifle revolutionized warfare and left an indelible mark on military strategies across the globe. Its impact can be examined through various lenses – technological advancements, its widespread adoption, and its role in insurgencies and guerrilla warfare.

The AK-47 introduced several technological advancements that significantly influenced subsequent firearm designs. It featured a gas-operated system with a rotating bolt, allowing for reliable functioning even in adverse conditions such as extreme temperatures or when exposed to dirt and debris. This innovation improved upon earlier designs prone to jamming or malfunctioning during combat. Additionally, it utilized a detachable magazine capable of holding thirty rounds – an impressive capacity at the time.

These features made the AK-47 more reliable and efficient than many contemporary firearms.

One of the key factors contributing to the AK-47's historical impact is its extensive adoption by military forces around the world. The Soviet Union actively distributed these rifles to their allies during the Cold War era, leading to their proliferation among communist states and liberation movements fighting against colonial powers. The affordability, ease of use, and robustness of this weapon made it particularly attractive for countries with limited resources or irregular armies.

Role in Insurgencies and Guerrilla Warfare:

The AK-47's impact on warfare extends beyond conventional battles between nation-states; it has played a pivotal role in insurgencies and guerrilla warfare throughout history. The weapon's simplicity allowed even untrained combatants to effectively engage opponents with minimal maintenance requirements. This accessibility empowered non-state actors such as rebel groups or insurgents to challenge well-equipped armies, leveling the playing field and changing the dynamics of conflicts.

The AK-47's ubiquity in conflicts spanning from Vietnam to Afghanistan and from Africa to Latin America demonstrates its enduring influence on asymmetric warfare.

Moreover, the psychological impact of the AK-47 cannot be understated. Its distinctive silhouette and reputation for reliability created an aura of fear among enemy combatants. This not only boosted the morale of those wielding it but also demoralized adversaries who faced a formidable opponent armed with a weapon that seemed unstoppable.

The AK-47 stands as a testament to Mikhail Kalashnikov's ingenuity and its profound impact on warfare. Its technological advancements, widespread adoption by military forces worldwide, and role in insurgencies have left an indelible mark on history. The AK-47 changed how wars were fought, empowering both state actors and non-state entities alike. As we reflect upon this iconic weapon, it is essential to understand not only its destructive power but also its lasting influence as a catalyst for military innovation and strategic adaptation across the globe.

***

Mikhail Kalashnikov, a Soviet weapons designer, is renowned for creating one of the most iconic firearms in history - the AK-47. However, his journey did not end with the creation of this revolutionary weapon; he faced numerous challenges and dedicated his life to improving upon his initial design. Through relentless determination and innovative thinking, Kalashnikov overcame these obstacles and left a lasting impact on the world of firearms.

One of the primary challenges Kalashnikov faced was enhancing the reliability and durability of his rifle. While the AK-47 had proven itself as a highly effective weapon, it still suffered from occasional malfunctions under extreme conditions. To address this issue, Kalashnikov introduced several modifications to strengthen its construction. He redesigned certain components, such as the receiver cover and bolt carrier group, using more robust materials that could withstand harsh environments without compromising performance.

Another significant challenge was reducing recoil while maintaining accuracy. The AK-47's design featured a gas-operated system that contributed to its simplicity but also generated substantial recoil forces upon firing. Recognizing the need for improved control during rapid fire or sustained engagements, Kalashnikov developed an innovative muzzle brake that effectively redirected gases to reduce felt recoil. This breakthrough not only enhanced shooter comfort but also allowed for more accurate follow-up shots.

Furthermore, ergonomics played a crucial role in enhancing user experience with firearms. The initial design of the AK-47 lacked certain ergonomic features that had become standard in modern rifles at the time. Realizing this limitation, Kalashnikov incorporated several improvements into subsequent iterations of his rifle. He introduced an adjustable stock length to accommodate different body sizes and shooting positions effectively. Additionally, he refined grip contours and redesigned controls for ease of use during stressful combat situations.

Kalashnikov's commitment to continuous improvement extended beyond ergonomics and reliability; he also sought to increase the weapon's versatility. Recognizing the changing nature of warfare, he developed different variants of the AK-47 to suit various combat scenarios. For example, he designed a shorter carbine version, known as the AKS-47, which was better suited for close-quarters combat or use by vehicle crews.

Additionally, he created specialized versions capable of mounting under-barrel grenade launchers or bayonets, further expanding their tactical capabilities.

Mikhail Kalashnikov faced numerous challenges throughout his journey of improving upon his initial design. From enhancing reliability and reducing recoil to refining ergonomics and increasing versatility, his innovations transformed the AK-47 into an even more formidable weapon. The modifications made by Kalashnikov have not only influenced subsequent generations of firearms but have also solidified his place as one of history's most influential weapons designers.

His unwavering dedication to refining his creation showcases the true spirit of innovation and has left an indelible mark on the world of firearms technology.

***

The AK-47, designed by Mikhail Kalashnikov, is one of the most iconic and widely recognized firearms in history. It has been praised for its reliability, simplicity, and effectiveness in combat. However, like any other invention of its magnitude, the rifle has not been without its controversies. In this section, we will delve into some of the criticisms and misconceptions surrounding the AK-47.

One common misconception is that the AK-47 is solely responsible for fueling conflicts worldwide. While it is true that this rifle has been used in numerous wars and armed conflicts over the past decades, it would be inaccurate to attribute these conflicts solely to this weapon. Political factors, socio-economic disparities, and historical grievances are typically at play when it comes to armed conflicts.

The AK-47 may have played a role due to its widespread availability and affordability; however, it should not shoulder all responsibility.

Another criticism often levied against the AK-47 is that it is a weapon of choice for terrorists and criminals worldwide. It is important to recognize that criminals or terrorists do not choose their weapons based on their merits or performance alone; instead, they consider factors such as accessibility and ease of use. The popularity of the AK-47 among these groups can be attributed more to its widespread availability rather than any inherent qualities that make it superior for unlawful activities.

One controversial aspect often discussed regarding Kalashnikov's creation is his own involvement in designing a weapon used for warfare. Critics argue that he should have considered the potential consequences before creating such a lethal firearm. However, it's crucial to understand that Kalashnikov was an engineer working within a specific context – Soviet military development during World War II – where creating an efficient weapon was seen as essential for national defense.

Moreover, some critics argue that despite its iconic status, the AK-47 is an outdated weapon that should be replaced by more modern firearms. They claim that its design and performance are inferior compared to contemporary rifles. While it is true that advancements in technology have led to the development of more sophisticated firearms, the AK-47's enduring popularity can be attributed to its simplicity and reliability, factors that make it well-suited for combat situations where maintenance and resources may be limited.

While the AK-47 has undeniably faced controversies throughout its history, it is important to critically analyze the criticisms and misconceptions surrounding this iconic rifle. It is not solely responsible for global conflicts or criminal activities but rather a tool utilized within complex socio-political contexts. Recognizing these nuances helps foster a more accurate understanding of this influential firearm's place in history.

***

The AK-47, designed by Mikhail Kalashnikov in the late 1940s, has become one of the most iconic and widely used firearms in history. Its global spread and influence can be attributed to various factors, including its reliability, affordability, ease of use, and widespread availability. This subtopic aims to explore the reasons behind the AK-47's proliferation worldwide and analyze its lasting impact on conflicts, society, and culture.

One key factor contributing to the global spread of the AK-47 is its remarkable reliability. The weapon's simplicity allowed it to function effectively even in extreme conditions with minimal maintenance. This crucial advantage made it highly appealing for armies operating in diverse environments ranging from deserts to jungles. The AK-47's ability to withstand rugged usage gave it a reputation as a battle-proven firearm that could be relied upon by soldiers worldwide.

Another significant reason for the rifle's proliferation was its affordability. Compared to other contemporary firearms of similar capabilities, the AK-47 was relatively inexpensive to produce due to its design simplicity. This factor enabled many countries with limited resources or emerging armed groups to acquire substantial quantities of these weapons at a fraction of the cost required for more advanced rifles. Consequently, this affordability allowed for wider access and distribution across various conflict zones.

The ease of use associated with the AK-47 also played a crucial role in its widespread adoption. Designed with ergonomics in mind, Kalashnikov crafted a weapon that could be operated effectively by both trained soldiers and inexperienced fighters alike. Its intuitive design made it accessible even for individuals without extensive training or technical knowledge, increasing its appeal among non-state actors who often lacked formal military training but sought effective tools for their operations.

Furthermore, throughout history, geopolitical dynamics have contributed significantly to the AK-47's proliferation. During periods characterized by Cold War tensions, both the Soviet Union and the United States fueled conflicts around the world by providing military aid to their respective allies. The AK-47 became a symbol of Soviet support for anti-colonial movements and communist regimes, leading to its adoption in numerous post-colonial struggles and proxy wars.

The global spread of the AK-47 has left an indelible mark on conflicts, societies, and cultures worldwide. Its presence in various regions has shaped warfare tactics and strategies, influencing battlefield dynamics for decades. Moreover, the weapon's symbolic value has permeated popular culture, appearing in movies, songs, literature, and art as a potent representation of rebellion or armed struggle.

The AK-47's global spread can be attributed to its reliability, affordability, ease of use, as well as geopolitical factors. Its impact on conflicts and societies worldwide is undeniable. As long as armed conflict persists across the globe or societies seek symbols of resistance or rebellion against oppressive forces—whether real or perceived—the AK-47 will continue to be an enduring icon of influence.

Mikhail Kalashnikov, the renowned Russian inventor and engineer, left an indelible mark on the world of weaponry. His most notable creation, the AK-47 assault rifle, revolutionized modern warfare and became a symbol of durability, reliability, and power. As a testament to his genius and contribution to the field of weaponry, Kalashnikov's legacy continues to be recognized and honored globally.

The AK-47's enduring popularity can be attributed not only to its simple design but also its exceptional performance in combat conditions. Its unparalleled reliability in adverse environments made it a weapon of choice for armies around the world. The rifle's success led to numerous variations and adaptations over the years, further solidifying its place as one of the most influential firearms in history.

Kalashnikov's contributions extended beyond just designing firearms; he also played a pivotal role in inspiring generations of weapon designers worldwide. His innovative approach emphasized simplicity without compromising effectiveness. This philosophy continues to influence firearm manufacturers today who strive to create weapons that are durable, easy to use, and capable of withstanding harsh conditions.

Furthermore, Kalashnikov's dedication towards improving soldiers' lives did not go unnoticed. He showed unwavering support for those who wielded his creations on battlefields by constantly working on enhancements that would enhance their safety and combat effectiveness. This commitment was recognized by numerous countries worldwide who bestowed upon him various awards and accolades throughout his lifetime.

In Russia itself, Kalashnikov achieved legendary status as a national hero due to his invaluable contributions. The Russian government honored him with numerous awards including the Order of St. Andrew - Russia's highest state decoration - for his exceptional service to his country. Additionally, several museums dedicated solely to showcasing his achievements were established across Russia as a tribute to his work.

Internationally too, Mikhail Kalashnikov's legacy has been widely celebrated. In 2004, the United Nations designated his birthday, November 10th, as "Kalashnikov Day" to commemorate his contributions to the field of weaponry. This recognition from a global platform further solidified his place in history.

The influence of Kalashnikov's designs extends far beyond military use. His firearms have become cultural icons and have made appearances in movies, television shows, and video games worldwide. The distinctive shape and reputation for reliability have turned the AK-47 into a symbol of resistance and revolution.

Mikhail Kalashnikov's contributions to weaponry continue to be recognized and celebrated globally. His creation of the AK-

47 not only transformed modern warfare but also inspired generations of weapon designers around the world. Through his dedication towards improving soldiers' lives and unwavering commitment to simplicity without compromising effectiveness, Kalashnikov became a national hero in Russia and gained international acclaim. The legacy he left behind is one that will forever be remembered as an exceptional testament to human ingenuity and innovation in the field of weapons design.

***

Mikhail Kalashnikov, a name synonymous with the invention of the world-renowned AK-47 assault rifle, holds a significant place in history as an icon. His creation not only revolutionized the field of firearms but also left a lasting impact on society that continues to be felt today.

Born in 1919 in a small village in Russia, Kalashnikov's early life was marked by hardship and adversity. However, it was during his time as a soldier during World War II that he developed a deep understanding of the need for reliable and efficient weaponry. This experience served as the catalyst for his groundbreaking invention.

The AK-47, or Avtomat Kalashnikova model 1947, became a symbol of power and resistance across the globe. Its simplicity in design and exceptional performance made it accessible to both trained soldiers and guerrilla fighters alike. This accessibility played a crucial role in shaping modern warfare strategies and conflicts worldwide.

Kalashnikov's creation had far-reaching effects beyond military applications. The AK-47 became prevalent not only on battlefields but also within various sociopolitical movements around the world. It became an emblem of rebellion and revolution against oppressive regimes, rendering its influence inseparable from numerous historical events.

One cannot discuss Kalashnikov's impact without acknowledging both its positive and negative aspects. On one hand, his invention provided defense to countless nations while ensuring their sovereignty against external threats. It enabled individuals to protect themselves against tyranny and gave oppressed communities hope for liberation.

On the other hand, the wide availability of AK-47 rifles led to unintended consequences that continue to plague societies today. The weapon's ease of use made it accessible to drug cartels, terrorists, and non-state actors who sought violence rather than protection or freedom. As such, it has become associated with international terrorism, insurgencies, and the illicit drug trade.

Despite the complex ethical questions surrounding his invention, Kalashnikov himself remained deeply proud of his creation. He viewed it as a tool of defense rather than aggression, emphasizing that he designed it to protect his homeland and its people. This perspective highlights the intricate relationship between intention and impact, reminding us of the multifaceted nature of technological advancements.

Mikhail Kalashnikov's legacy extends beyond the realm of firearms. His contributions to engineering and design continue to inspire innovation in various fields. Moreover, his story serves as a reminder that even seemingly small inventions can have profound consequences for society.

As we reflect on Mikhail Kalashnikov's lasting impact on society, we must acknowledge both the positive and negative aspects associated with his invention. The AK-47 revolutionized warfare while becoming an emblem of resistance against oppression. However, its widespread availability also contributed to violence and illicit activities by terrorist groups worldwide. It is through this nuanced understanding that we can truly appreciate the complexities surrounding this iconic figure and his creation, prompting us to question the broader implications of technological advancements for our society as a whole.

Eugene Stoner

1922-1997

Eugene Morrison Stoner, an iconic figure in the world of firearms design, revolutionized the industry with his innovative creations. Born on November 22, 1922, in Gosport, Indiana, Stoner's relentless pursuit of excellence and commitment to pushing boundaries led him to develop some of the most influential firearms in history. From his humble beginnings as a self-taught engineer to his role as the mastermind behind legendary rifles such as the AR-15 and M16, Stoner's impact on modern warfare and civilian gun ownership cannot be overstated.

Stoner's interest in engineering and mechanics began at an early age. As a teenager, he would spend countless hours tinkering with various machines and gadgets, driven by an insatiable curiosity about how things worked. Despite lacking formal education in engineering or firearms design, Stoner possessed an innate talent for innovation that would set him apart from his peers.

In 1955, Stoner joined ArmaLite - a small division of Fairchild Engine and Airplane Corporation - where he would make history. Tasked with designing a lightweight rifle capable of firing intermediate cartridges for military use, he embarked on what would become his most significant achievement: the AR-15. This groundbreaking weapon featured several groundbreaking design elements that provided enhanced accuracy, reliability, and ease of use.

The AR-15's unique gas-operated system allowed for reduced recoil and improved control during rapid fire sequences – a crucial advantage on the battlefield. By incorporating lightweight materials such as aluminum alloys and advanced polymers into its construction, Stoner created a rifle that was not only more maneuverable but also easier to maintain than its predecessors.

However innovative it may have been initially; it was not until later developments when it truly gained worldwide recognition. In 1959, after Colt acquired ArmaLite's assets – including the AR-15 design – they refined and rebranded it as the M16. Subsequently, the M16 became the standard-issue rifle for U.S. troops during the Vietnam War, forever solidifying Stoner's reputation as a visionary firearms designer.

Stoner's contributions extended beyond military applications. The AR-15 platform, now widely known as the modern sporting rifle (MSR), became immensely popular among civilian shooters due to its versatility and customization options. Its lightweight construction and adaptable design made it a favorite among sportsmen, hunters, and firearms enthusiasts around the world.

Eugene Morrison Stoner's legacy is one of innovation and adaptability that continues to shape the firearms industry today. His pioneering designs set new standards for reliability, efficiency, and ergonomics in both military and civilian firearm development. Whether through his groundbreaking AR-15/M16 or his numerous other contributions to small arms design, Stoner's impact on American firearm culture remains immeasurable.

In this comprehensive exploration of Eugene Morrison Stoner's life and work, we will delve into his groundbreaking designs, examine their impact on modern warfare and civilian gun ownership, as well as uncover lesser-known aspects of this legendary American firearms designer.

***

Eugene Stoner, born on November 22, 1922, in Gosport, Indiana, was an American firearms designer and engineer who is best known for his revolutionary contributions to the development of firearms. His creations laid the foundation for many modern military rifles and forever changed the landscape of small arms.

Stoner's early life was marked by a strong interest in mechanics and engineering. Growing up in a rural farming community, he developed a deep appreciation for machinery and spent countless hours tinkering with various devices. This passion led him to pursue an education that would further develop his skills and knowledge.

After completing high school, Stoner enrolled at the Worcester Polytechnic Institute (WPI) in Massachusetts. At WPI, he studied mechanical engineering with a focus on machine design. It was during this time that he honed his technical abilities and gained valuable insights into the principles of firearm mechanics.

While studying at WPI, World War II erupted, prompting Stoner to put his education on hold temporarily. In 1942, he enlisted in the United States Marine Corps and served as an aviation ordnance technician during the war. This experience provided him with practical exposure to firearms and ammunition systems used by the military at the time.

Upon returning from his military service in 1945, Stoner resumed his studies at WPI. Determined to apply what he had learned to real-world applications, he sought out opportunities for practical experience beyond academic settings. He secured internships at various manufacturing companies where he gained hands-on experience working with cutting-edge machinery.

In 1954, Eugene Stoner's career took a significant turn when he joined Armalite Corporation as their chief engineer. Armalite was a small division of Fairchild Engine & Airplane Corporation specializing in small arms development. This collaboration proved pivotal as it provided Stoner with an environment conducive to innovation and experimentation.

During his tenure at Armalite, Stoner was tasked with developing a lightweight and reliable rifle for the military. His groundbreaking work led to the creation of the AR-10 rifle, which incorporated several innovative features such as a gas-operated system and lightweight materials. While the AR-10 did not gain immediate traction with the military, it laid the foundation for Stoner's most iconic creation – the AR-15.

Stoner's relentless pursuit of firearm innovation and his dedication to pushing boundaries led him to refine his design further. In 1956, he introduced the AR-15 rifle, which would later become widely adopted by various military forces around the world as the M16.

Eugene Stoner's early life and education played a crucial role in shaping him into an exceptional firearms designer. His passion for mechanics, combined with his formal education and practical experience, laid a solid foundation for his future accomplishments. The stage was set for him to revolutionize small arms technology and leave an indelible mark on modern warfare.

Eugene Morrison Stoner, an American firearms designer, is renowned for his groundbreaking contributions to the field of small arms. His most notable creation, the Stoner Rifle, revolutionized modern warfare and became the foundation for numerous military rifles used around the world. To truly understand the genius behind this invention, it is essential to delve into Eugene Stoner's innovative mindset and explore the genesis of his revolutionary rifle.

Stoner's journey as a firearms designer began during World War II when he worked at a machine shop producing machine guns for the war effort. This experience sparked his interest in firearm design and set him on a path towards innovation. After joining Armalite Corporation in 1955, he dedicated himself to creating a lightweight rifle that would challenge traditional conventions.

One of Stoner's key insights was recognizing that conventional infantry rifles were excessively heavy and outdated. He believed that reducing weight without compromising effectiveness was crucial for future military engagements. This realization fueled his determination to develop a firearm that would address these concerns.

Stoner's first major breakthrough came with the creation of the AR-10 rifle in 1956. Unlike its predecessors, this rifle utilized lightweight materials such as aluminum alloys and polymers instead of traditional steel components. By employing synthetic materials intelligently, Stoner achieved significant weight reduction without sacrificing durability or accuracy.

However, it was with the subsequent development of the AR-15 where Eugene Stoner truly showcased his innovative mindset. Recognizing that conventional ammunition designs limited firepower and range capabilities, he explored alternative solutions. He conceptualized an intermediate cartridge that combined aspects of both traditional pistol and full-sized rifle rounds – resulting in what we now know as .223 Remington or 5.56x45mm NATO caliber.

This cartridge facilitated increased accuracy at longer ranges while maintaining manageable recoil compared to larger calibers used by standard infantry rifles at the time. Stoner's forward-thinking approach to ammunition design allowed for higher ammunition capacity and reduced weight, further enhancing the rifle's overall performance.

Moreover, Stoner incorporated a gas-operated, direct impingement system into the AR-15. This innovative mechanism eliminated the need for a piston-driven system and significantly reduced recoil. By redirecting high-pressure gas from fired rounds directly into the firearm's bolt carrier group, Stoner achieved enhanced reliability and ease of maintenance.

The culmination of these groundbreaking advancements led to the creation of what would become one of the most successful rifles in history – the M16. Adopted by the United States military in 1964, this rifle was a direct descendant of Eugene Stoner's original designs. Its lightweight construction, improved accuracy, and increased firepower revolutionized modern infantry tactics and influenced subsequent rifle designs worldwide.

Eugene Stoner's innovative mindset enabled him to challenge conventional norms and revolutionize small arms design. Through his persistence, he conceived a lightweight rifle that combined cutting-edge materials with advanced ammunition technology – forever changing the face of modern warfare. The genesis of the Stoner Rifle stands as a testament to his visionary thinking and continues to shape firearms development today.

***

In the realm of firearms, few names carry as much weight as Eugene Stoner. Renowned for his groundbreaking work in firearm design, Stoner's most notable contribution came in the form of the AR-15 rifle. Widely regarded as one of the most influential firearms in history, the AR-15 revolutionized firearm design and left an indelible mark on both military and civilian use.

Eugene Stoner's journey to creating the AR-15 began in the late 1950s when he joined Armalite, a small firearms company.

Tasked with designing a lightweight and reliable rifle for military use, Stoner set out to challenge conventional firearm designs. His innovative approach led to the creation of what would later become known as the AR-15. One of Stoner's key breakthroughs was his use of lightweight materials such as aluminum alloys and synthetic polymers.

By incorporating these materials into the rifle's construction, he significantly reduced its weight without compromising its structural integrity or durability. This made it easier for soldiers to maneuver and carry their weapons over long distances without sacrificing firepower. Another pivotal aspect of Stoner's design was his utilization of a gas-operated, direct impingement system. Unlike traditional piston-driven systems that required additional moving parts, this system utilized high-pressure gas from fired cartridges to cycle the weapon automatically.

Not only did this simplify maintenance requirements by eliminating unnecessary components, but it also increased reliability and reduced recoil – factors crucial for accurate rapid-fire shooting. Stoner's attention to ergonomics played a crucial role in making the AR-15 intuitive and comfortable to handle. He introduced features such as a pistol grip that allowed for a more natural grip angle while enhancing control during firing sequences.

Additionally, he incorporated adjustable stocks that catered to different body sizes and shooting positions – an innovation that further improved user experience. The AR-15's adaptability was yet another aspect that set it apart. Stoner designed the rifle to be modular, allowing for easy customization and compatibility with various accessories. This flexibility meant that users could modify their firearms to suit their specific needs, whether it be attaching optics, suppressors, or different barrel lengths.

This adaptability made the AR-15 a highly versatile weapon platform suitable for a wide range of applications. Stoner's groundbreaking work on the AR-15 did not go unnoticed. In 1963, the United States military adopted an adapted version of his design as the M16 rifle. The M16 went on to become one of the most widely used rifles in modern warfare and served as a testament to Stoner's ingenuity.

Beyond its military applications, the AR-15 found immense popularity among civilian shooters due to its exceptional performance and versatility. Today, countless variants and clones of Stoner's original design are available on the market, making it one of the most popular firearms in civilian ownership. Eugene Stoner's innovative approach to firearm design forever changed the landscape of modern weaponry with his creation of the AR-15 rifle.

***

The AR-15, a rifle that has become synonymous with American gun culture, has an intriguing history that traces its roots back to military adoption and subsequently gaining immense popularity among civilians. Developed by Eugene Stoner in the late 1950s, the AR-15 was originally intended for military use but soon found its way into civilian hands, becoming one of the most popular firearms in the United States.

Eugene Stoner, an American firearms designer and engineer, created the AR-15 while working for ArmaLite, a small arms manufacturing company. The weapon was initially designed as a lightweight and versatile firearm for military purposes. In 1959, ArmaLite sold the rights to produce the AR-15 to Colt's Manufacturing Company. Colt recognized its potential and began marketing it to various branches of the U.S. military.

The AR-15's journey from military adoption to civilian popularity began during the Vietnam War era when it was first deployed by American troops in combat zones. Soldiers appreciated its lightweight design and accuracy, making it an ideal weapon for close quarters combat situations. As more servicemen used and praised it on the battlefield, awareness of this exceptional rifle spread among civilians back home.

Following their return from Vietnam, many veterans sought to own firearms resembling those they had used during their service. This desire led them towards purchasing surplus or civilian versions of military rifles like the AR-15. However, due to certain legal restrictions at that time, civilians were not allowed access to fully automatic versions of these rifles.

In response to this demand from civilians seeking an alternative option similar to their beloved military rifles without fully automatic capabilities, manufacturers introduced semi-automatic variants of these guns into the market. These semi-automatic versions retained all other features, but limited firing rates solely based on individual trigger pulls.

The introduction of semi-automatic versions opened up a new avenue for the AR-15's civilian popularity. Its lightweight design, customizable features, and ease of use made it an attractive choice for various purposes such as hunting, sport shooting, and home defense. Over time, the AR-15's reputation grew exponentially among gun enthusiasts due to its reliability and versatility.

The AR-15's rise in popularity was further fueled by its accessibility and widespread availability across the United States. It became a common sight at shooting ranges, competitions, and even in movies and television shows. The rifle's adaptability to different calibers and the ability to attach various accessories such as optics, grips, and lights made it highly sought after by firearms enthusiasts who desired personalized modifications.

Today, the AR-15 remains one of the most popular rifles in America. Its journey from military adoption to civilian popularity showcases how a firearm designed for military purposes can capture the imagination of civilians seeking reliable self-defense weapons or those with a passion for recreational shooting activities. While debates surrounding its use continue, there is no denying that Eugene Stoner's creation has left an indelible mark on American gun culture.

***

When discussing the most influential firearms in modern warfare, the name Eugene Stoner and his iconic creation, the M16 rifle, cannot be overlooked. The M16 revolutionized military small arms with its innovative design and remarkable performance. Developed during the 1950s and 1960s, this weapon became synonymous with American military might and forever changed the landscape of warfare.

Eugene Stoner, an American firearms designer, played a pivotal role in developing numerous groundbreaking firearms throughout his career. However, it was his work on the M16 that solidified his place in history as one of the greatest firearm designers of all time. The rifle's lightweight construction and advanced engineering made it a game-changer on the battlefield.

One of the key features that set the M16 apart from its predecessors was its use of lightweight materials such as aluminum and composite plastics. This reduced its weight significantly compared to traditional steel rifles, making it easier for soldiers to carry during extended operations. Additionally, this lightweight construction allowed for better maneuverability in tight spaces and improved accuracy when firing on-the-move.

Stoner's design also incorporated a gas-operated system that utilized direct impingement technology. This innovation eliminated conventional piston systems found in other rifles at that time. By redirecting propellant gases directly into the bolt carrier group to cycle ammunition, Stoner achieved reduced recoil and improved reliability—a significant advantage for soldiers engaged in combat situations.

Furthermore, Stoner's design introduced a unique detachable box magazine that could hold up to thirty rounds of ammunition—a substantial increase compared to previous rifles' capacities. This higher ammunition capacity allowed soldiers to sustain firefights for longer durations without needing frequent reloads—a critical factor when engaging enemies on ever-evolving battlefields.

Another notable contribution by Stoner was the development of smaller caliber .223 Remington/5.56mm NATO rounds specifically designed for the M16. These rounds were lighter and faster, yet still lethal, further enhancing the rifle's effectiveness. The smaller caliber meant soldiers could carry more ammunition while maintaining accuracy and range.

The adoption of the M16 by the United States military in the 1960s marked a turning point in modern warfare. Its lightweight design, reliability, increased ammunition capacity, and improved accuracy revolutionized infantry tactics. Soldiers armed with this rifle could engage targets at greater distances while maintaining maneuverability—a significant advantage on the battlefield.

Throughout its service life, the M16 platform has continued to evolve and adapt to meet changing needs. Various iterations have been developed for different military branches and specialized units, all based on Eugene Stoner's original design principles.

Eugene Stoner's contribution to modern warfare through his creation of the M16 rifle cannot be overstated. His innovative design concepts transformed military small arms forever. The lightweight construction, gas-operated system, detachable box magazine, and smaller caliber rounds all contributed to making the M16 an iconic weapon that set new standards for performance on the battlefield. Its impact continues to be felt today as it remains one of the most widely used rifles in armed forces around the world.

***

The AR-15 and its military counterpart, the M16, are iconic rifles that revolutionized firearm design. Developed by Eugene

Stoner in the late 1950s, these rifles have had a significant impact on both military engagements and civilian firearms culture. However, their history is not without controversies, ranging from reliability issues to being demonized by anti-gun groups.

One of the earliest controversies regarding these rifles was their reliability during combat operations in Vietnam. When first introduced to U.S. troops in 1963, the M16 faced severe reliability issues due to ammunition and maintenance problems. Soldiers reported frequent malfunctions, such as jamming, which put them at risk during crucial moments in battle. The unreliability of these rifles led to a loss of confidence among soldiers and sparked debates about their effectiveness on the battlefield.

Another highly debated controversy revolves around the civilian availability of AR-15 rifles and their involvement in mass shootings. Due to its semi-automatic nature and modular design that allows for customization with various accessories, critics argue that it has become a weapon of choice for individuals seeking to commit heinous acts of violence. Although long guns are seldom used in murders, it hasn't stopped Bloomberg-led anti-gun groups from trying to convince the American public that these rifles are evil.

Furthermore, controversy surrounds the use of certain features associated with AR-15-style rifles, such as detachable magazines capable of holding more than ten rounds. Critics argue that these features enhance a shooter's ability to cause harm quickly without reloading frequently, making them particularly dangerous in mass shooting scenarios. Once again, this argument fails. The Ruger Mini-14 fires the same cartridge and can take a magazine, but it doesn't look as scary. Many anti-gun advocates are okay with that firearm because it doesn't look scary.

Additionally, the controversy surrounding the AR-15 and M16 rifles extends to their use by law enforcement agencies. Some argue that these rifles contribute to the militarization of police forces, as they believe that the guns are designed for combat situations rather than typical domestic policing. Critics assert that this militarization can lead to an escalation of force and undermine community trust in law enforcement.

Despite these controversies, it is important to note that many gun enthusiasts and supporters of the Second Amendment value the AR-15 and M16 rifles for their versatility, accuracy, and modularity. Advocates argue that blaming firearms themselves overlooks other underlying factors contributing to violence in society and emphasize responsible ownership as a crucial aspect of firearm safety.

Eugene Morrison Stoner was a pioneering American firearms designer whose contributions revolutionized the field of small arms. His innovative designs, including the iconic AR-15 rifle, have left an indelible mark on military weaponry and continue to shape modern firearms technology. Today, his legacy is celebrated, and his contributions are recognized by enthusiasts, scholars, and military professionals worldwide.

Stoner's most significant contribution was undoubtedly the creation of the AR-15 rifle. Originally designed in the late 1950s as a lightweight alternative to traditional battle rifles, the AR-15 featured a gas-operated system that reduced recoil and improved accuracy. Its modular design allowed for easy customization and adaptation to various combat scenarios. The AR-15's success led to its adoption by the United States military as the M16 rifle, which became a standard issue during the Vietnam War.

The impact of Stoner's design went far beyond its military applications. The civilian version of the AR-15, known as the semi-automatic "AR-15 style" rifle, became immensely popular among gun enthusiasts for its versatility and ease of use. Today, it remains one of the most widely owned firearms in America.

Stoner's ingenuity extended beyond his work on rifles. He also played a pivotal role in developing other firearms systems that showcased his engineering brilliance. Notably, he designed the Stoner 63 modular weapon system—a versatile family of firearms that could be quickly reconfigured for different roles, such as assault rifles or light machine guns—providing soldiers with adaptability on the battlefield.

In recognition of his exceptional contributions to firearm design, Eugene Morrison Stoner received numerous accolades throughout his career. In 1991, he was inducted into both the National Firearms Museum Hall of Fame and The Sporting Goods Industry Hall of Fame—an honor bestowed upon individuals who have made significant advancements in their respective fields. These inductions not only recognized Stoner's technical achievements but also highlighted his impact on the broader firearms industry.

Moreover, Stoner's legacy lives on through the continued use and development of his designs. Many modern military and law enforcement agencies worldwide still employ variants of the AR-15 or its derivatives. The rifle's ongoing success is a testament to Stoner's vision and the enduring relevance of his designs in contemporary combat scenarios.

Beyond military applications, Eugene Morrison Stoner's impact can be seen in technological advancements within the firearms industry. His modular design concepts have influenced subsequent generations of firearms, leading to innovations such as quick-change barrels and accessory rails that allow for easy attachment of optics and other tactical equipment.

Eugene Morrison Stoner's contributions to firearm design have left an indelible mark on both military weaponry and civilian gun culture. His revolutionary designs continue to shape modern small arms technology, with the AR-15 remaining a symbol of innovation and adaptability. Through recognition in various halls of fame and the ongoing use of his designs by armed forces worldwide, Stoner's legacy as one of history's most influential firearms designers is firmly established.

The name Eugene Stoner is synonymous with revolutionizing the American firearms industry. His innovative designs and engineering prowess have left an indelible mark on the way firearms are manufactured and used in the United States. From his groundbreaking creation of the AR-15 rifle to his contributions to military weaponry, Stoner's impact continues to shape the landscape of American firearms.

At the heart of Stoner's influence is his most famous invention, the AR-15 rifle, which later evolved into the popular civilian variant, the M16. Introduced in 1956, this lightweight semi-automatic rifle revolutionized modern warfare and transformed how soldiers engaged in combat. Its design incorporated lightweight materials and a gas-operated system that significantly reduced recoil, making it more accurate and controllable than its predecessors.

The AR-15's adaptable nature also allowed for easy customization, catering to individual preferences and mission requirements.

Stoner's AR-15 design not only gained recognition within military circles but also caught the attention of civilians who sought similar capabilities for sporting and self-defense purposes. The commercial success of this platform paved the way for numerous manufacturers to produce their own versions of this iconic firearm. Today, it remains one of America's most popular rifles due to its versatility, reliability, and ease of use.

Beyond his contributions to small arms design, Stoner was instrumental in developing other significant military weapons systems. His work on light machine guns, such as the M60, helped increase mobility while maintaining firepower on battlefields during World War II and subsequent conflicts. Additionally, he played a crucial role in designing grenade launchers like the M79 and M203, which enhanced soldiers' ability to engage enemies at longer ranges.

Stoner's impact extended beyond individual weapons; he fundamentally altered manufacturing techniques within the firearms industry as well. By introducing synthetic materials like polymer into firearm construction instead of traditional wood or metal components, he significantly reduced weight while maintaining durability. This shift not only improved the ergonomics of firearms but also made them more cost-effective to produce, opening up opportunities for wider accessibility.

Furthermore, Stoner's innovative approach to firearm design spurred a culture of innovation within the American firearms industry. His relentless pursuit of perfection and willingness to challenge conventional wisdom inspired subsequent generations of engineers and designers. Today, his influence can be seen in the wide array of modern firearms that continue to evolve and push the boundaries of technology and performance.

Eugene Stoner's impact on the American firearms industry cannot be overstated. His revolutionary designs, particularly the AR-15 rifle and its variants, have redefined small arms capabilities both in military and civilian settings. Through his ingenuity and forward-thinking approach, Stoner forever changed how firearms are designed, manufactured, and utilized in the United States. His legacy lives on through countless innovations that continue to shape the industry today.

Eugene Morrison Stoner, an American firearms designer, left an indelible mark on the world with his innovative creations and revolutionary ideas. Throughout his career, he challenged conventional norms and pushed boundaries in the field of small arms design. From the iconic AR-15 rifle to his groundbreaking concepts like modular weapon systems, Stoner's contributions have had a lasting impact on both military and civilian firearm development.

One cannot discuss Stoner's influence without acknowledging his most renowned creation, the AR-15 rifle. Initially developed as a lightweight and versatile weapon for military use, it quickly gained popularity among civilians as well. Its modular design allowed for various accessories and attachments to be easily integrated, making it customizable to individual preferences. This adaptability appealed to law enforcement agencies, sports shooters, and hunters alike.

Stoner's forward-thinking approach revolutionized small arms design by introducing modern materials like aluminum alloys and polymers into firearm construction. By replacing traditional wood stocks with lightweight synthetic materials, he significantly reduced the weight of rifles while maintaining durability. This breakthrough not only enhanced maneuverability but also influenced future designs across industries.

Furthermore, one cannot overlook Stoner's visionary concept of modular weapon systems. His idea centered around creating firearm platforms that could be easily modified based on specific mission requirements or user preferences. This notion laid the foundation for future advancements in firearms technology, such as interchangeable barrels or caliber conversions, with minimal effort.

Stoner's influence extended beyond mere technical innovations; it also encompassed a broader understanding of ergonomics and user experience. He emphasized ease of use, reliability under adverse conditions, and minimal recoil – factors that continue to shape contemporary firearm designs today.

In terms of military applications, Stoner's creations played a crucial role in shaping modern warfare tactics by providing soldiers with lighter weaponry that increased their mobility without sacrificing firepower. The introduction of the M16 rifle during the Vietnam War marked a turning point in small arms development, as it replaced the heavier and less reliable M14 rifle. Stoner's design not only improved soldiers' efficiency on the battlefield but also influenced future generations of military firearms.

Moreover, Stoner's impact transcends his specific creations. His commitment to innovation, attention to detail, and dedication to pushing boundaries have inspired countless designers and engineers in various fields. His legacy lives on through the continued exploration of new materials, designs, and concepts that aim to improve firearm performance and user experience.

Eugene Morrison Stoner's enduring influence on small arms design cannot be overstated. From his groundbreaking AR-15 rifle to his innovative ideas surrounding modular weapon systems, Stoner pushed boundaries and challenged traditional norms. His contributions have shaped modern firearms technology by introducing lightweight materials, improving ergonomics, and emphasizing adaptability. Furthermore, his impact extends beyond the realm of firearms alone – he has inspired generations of designers with his relentless pursuit of innovation.

Eugene Morrison Stoner will forever be remembered as a visionary whose ideas continue to shape the future of small arms design.

Gaston Glock
1929-

In the world of firearms, few names carry as much weight as Gaston Glock. With his unwavering dedication to innovation and relentless pursuit of perfection, Glock has revolutionized the industry and created what is now known as the world's most popular handgun. His story is one of ambition, determination, and an unyielding commitment to excellence.

Born in 1929 in Vienna, Austria, Gaston Glock initially had no connection to the world of firearms. In fact, he began his career as a successful engineer specializing in advanced polymers and synthetic materials. It was this expertise that eventually led him down a path that would forever change the landscape of firearm design.

In the early 1980s, Austria's military issued a call for tenders for a new service pistol. At that time, most handguns were made with traditional materials such as steel and wood. However, Glock saw an opportunity to apply his knowledge of polymers to create something entirely new – a lightweight and durable handgun that would surpass all existing designs.

Undeterred by his lack of experience in firearms manufacturing, Glock embarked on an ambitious endeavor to develop a revolutionary pistol from scratch. Drawing on his engineering background and collaborating with experts from various fields such as metallurgy and ergonomics, he set out to create a firearm that would be reliable under any circumstances.

After years of research and development, Gaston Glock unveiled his first creation – the Glock 17 – in 1982. The pistol featured several groundbreaking design elements that set it apart from its competitors. One notable innovation was its use of high-strength polymer frames instead of traditional metal frames. This not only reduced weight but also improved durability while maintaining exceptional performance.

What truly differentiated Glock's pistols was their simplicity and reliability. Unlike many conventional handguns at the time, which required numerous external safeties or complex mechanisms for operation, Glock pistols employed a unique "safe action" system. This innovative design integrated three independent safeties into the trigger mechanism, enabling users to confidently carry their firearms with reduced risk of accidental discharge.

The Glock 17 quickly gained recognition for its exceptional performance and reliability. It was adopted by numerous law enforcement agencies around the world, including the Austrian military and police. Its popularity among professionals soon spilled over into civilian markets, making it one of the most sought-after handguns by firearm enthusiasts worldwide.

Today, more than three decades after its introduction, Glock remains at the forefront of firearms manufacturing. Gaston Glock's relentless pursuit of perfection and his commitment to producing reliable and high-performance handguns have solidified his legacy as an industry icon.

In this article, we will delve deeper into Gaston Glock's life and explore how his passion for innovation and dedication to excellence have forever changed the landscape of firearm design. Join us as we uncover the remarkable journey of a man who revolutionized an industry through his unwavering determination to create the world's most popular handgun.

***

Firearms have played a significant role in human history, evolving from simple hand cannons to highly advanced weapons. Throughout the centuries, numerous inventors and innovators have contributed to the development of firearms technology. However, it was Gaston Glock who revolutionized the industry with his groundbreaking design that would become the world's most popular handgun - the Glock pistol.

To understand how Glock pistols changed the game, it is essential to delve into a brief history of firearms technology. The first handheld firearms emerged in China during the 13th century, known as "fire lances." These early devices were essentially tubes filled with gunpowder and projectiles, ignited by a lit fuse. Over time, these primitive weapons evolved into more sophisticated firearms like matchlock muskets and flintlock pistols.

The 19th century witnessed tremendous advancements in firearms technology. Innovations such as percussion caps and rifling significantly improved accuracy and reliability. The introduction of revolvers and repeating rifles further transformed warfare and self-defense capabilities.

Fast forward to the late 20th century when Gaston Glock entered the scene. A relatively unknown Austrian engineer with no prior experience in firearm design, he set out to create a handgun that would surpass all others in terms of reliability, simplicity, and performance.

Glock's breakthrough came with his innovative use of polymers instead of traditional steel frames for handgun construction. This allowed for lighter weight without compromising strength or durability - a significant departure from conventional designs at that time. Additionally, Glock introduced an unconventional striker-fired mechanism instead of relying on external hammers or internal firing pins like most handguns.

The culmination of these innovations was unveiled in 1982 with the introduction of the Glock 17 - a semi-automatic pistol chambered for 9mm ammunition. It quickly gained popularity among law enforcement agencies due to its high-capacity magazine (holding up to 17 rounds) and its remarkable reliability, even under adverse conditions.

The Glock's success continued to grow as it gained a reputation for being virtually indestructible and having minimal maintenance requirements. Its simple design made it easier to disassemble and clean, making it an ideal choice for both professional shooters and novice gun owners alike.

The Glock's impact on the firearms industry was undeniable. Its influence prompted other manufacturers to adopt polymer frames, leading to a paradigm shift in handgun design. The Glocks' ergonomic grip, ambidextrous controls, and customizable options further cemented their popularity among shooters of all skill levels.

Today, Glock pistols are ubiquitous in both civilian and military markets worldwide. Their reliability, simplicity, and performance have made them the gold standard for handguns.

Gaston Glock's vision of creating a firearm that would change the game has undoubtedly been realized with his revolutionary pistol design.

The history of firearms technology has seen numerous advancements over centuries. However, it was Gaston Glock's innovative use of polymers and striker-fired mechanisms that revolutionized the industry with his iconic handgun designs. The unmatched reliability and simplicity of Glock pistols have forever changed the game, establishing them as the most popular handguns globally.

***

In the world of firearms, few names are as recognizable and revered as Glock. Synonymous with reliability, durability, and innovation, Glock pistols have become the weapon of choice for law enforcement agencies, military personnel, and gun enthusiasts worldwide. Behind this iconic brand stands one man - Gaston Glock - who revolutionized the industry and built a global empire on semi-automatic pistols.

Born in 1929 in Austria, Gaston Glock originally had no connection to the firearms industry. He was an engineer specializing in polymers and plastics. However, in the early 1980s, when Austria's Ministry of Defense announced its search for a new military sidearm to replace their aging pistols, fate intervened. Unburdened by preconceived notions about firearm design, Gaston decided to take on this challenge despite having no prior experience in gun manufacturing.

Gaston assembled a team of experts from various fields to help him design a pistol that would surpass all existing standards. Drawing inspiration from his background in polymer engineering, he incorporated advanced materials like polymer frames into his designs. This not only made the guns lighter but also significantly increased their resistance to corrosion and wear.

In 1982, after countless prototypes and rigorous testing procedures set by Austrian authorities, Gaston's creation - the Glock 17 - emerged victorious over established manufacturers such as Beretta and Sig Sauer. The Austrian military adopted it as their official service pistol.

This triumph marked the beginning of an unprecedented rise for both Glock and its founder. The combination of innovative design features such as striker-fired mechanisms instead of traditional hammer firing systems coupled with exceptional reliability quickly garnered attention from law enforcement agencies across the globe.

Gaston's relentless pursuit of perfection led to continuous improvements in his firearms' performance and safety features over time. This commitment contributed immensely to creating trust among users who relied on Glock pistols in life-or-death situations. The company's unwavering dedication to customer satisfaction, coupled with their commitment to producing reliable and affordable firearms, fueled their exponential growth.

As word spread about the Glock's reliability and ease of use, demand skyrocketed. Gaston Glock expanded his manufacturing capabilities to meet the burgeoning market needs. The decision to establish a manufacturing facility in the United States further solidified Glock's dominance in the global firearms industry.

Today, Glock pistols are sold in over 100 countries and have become a staple choice for military personnel, law enforcement agencies, and civilians alike. Gaston Glock's vision of designing a dependable handgun that would surpass all expectations has not only transformed his own life but also revolutionized the industry at large.

Gaston Glock's journey from an engineer specializing in polymers to becoming the founder of one of the world's most popular handgun brands is truly remarkable. His relentless pursuit of excellence and innovative approach towards firearm design have propelled him from obscurity into legendary status within the firearms community. With each passing year, the rise of Glock continues as Gaston's legacy lives on through his revolutionary semi-automatic pistols.

***

In the realm of firearms, one name stands above the rest – Gaston Glock. His revolutionary design for a polymer-framed handgun not only transformed the industry but also became an iconic symbol of reliability and innovation. The birth of the Glock handgun can be traced back to a time when traditional steel-framed pistols dominated the market.

It was in the early 1980s when Gaston Glock, an Austrian engineer with no prior experience in firearms manufacturing, received an invitation from his country's military to submit a proposal for a new standard-issue sidearm. This unexpected opportunity set in motion a chain of events that would forever change the world of handguns.

Glock approached this challenge with fresh eyes and an open mind, unburdened by preconceived notions about firearm design. His engineering background allowed him to approach the problem from unconventional angles, ultimately leading to breakthrough innovations that would define his brand.

One key aspect that set Glock's design apart was its use of polymer materials instead of traditional steel frames. At that time, polymers were primarily associated with low-quality consumer products rather than high-performance firearms. However, Glock saw potential in this lightweight and corrosion-resistant material and decided to incorporate it into his handgun design.

Another groundbreaking feature was Glock's introduction of a striker-fired mechanism instead of a traditional hammer-fired system. This innovation removed unnecessary complexity and reduced moving parts, enhancing reliability while simplifying maintenance requirements. The striker-fired system provided consistent trigger pull characteristics shot after shot, ensuring greater accuracy for shooters.

Furthermore, Gaston Glock implemented several ergonomic features into his design to improve user experience. The pistol featured textured grips for enhanced control and comfort during firing, while its low bore axis reduced muzzle rise and improved recoil management – characteristics rarely seen before on handguns.

When it came time to unveil his creation – now known as the "Glock 17" due to its 17-round magazine capacity – Gaston Glock astounded the firearms community with his revolutionary design. The Glock 17's lightweight construction, superior reliability, and ease of use captivated military forces, law enforcement agencies, and civilian shooters alike.

The Glock handgun's debut marked a turning point in firearm design history. Its success challenged the dominance of traditional steel-framed pistols and shattered long-standing prejudices against polymer materials. Gaston Glock had not only created a highly functional firearm but had also revolutionized the industry by introducing innovative manufacturing processes.

Today, the Glock brand has become synonymous with quality, reliability, and innovation. The company continues to refine its designs based on user feedback and technological advancements while maintaining the core principles that made their handguns so successful – durability, simplicity, and performance.

Gaston Glock's pursuit of excellence led to the birth of an iconic handgun that forever changed the industry. His revolutionary design incorporating polymer frames, striker-fired mechanisms, ergonomic enhancements, and innovative manufacturing processes set new standards for reliability and functionality. The Glock handgun remains a testament to one man's vision and determination to challenge conventionality in pursuit of perfection.

***

When it comes to firearms, one name that stands out above the rest is Gaston Glock. His revolutionary creation, the Glock pistol, has not only become the world's most popular handgun but has also redefined firearm safety and reliability. By challenging conventional norms and breaking barriers, Glock pistols have set new standards in the industry.

One of the key aspects that sets Glock pistols apart is their innovative design. Unlike traditional handguns that relied on a single-action or double-action firing mechanism, Gaston Glock introduced a striker-fired system. This design eliminates the need for an external hammer and provides a consistent trigger pull with each shot. This breakthrough not only simplifies the operation of the firearm but also enhances safety by reducing accidental discharges caused by snagging on clothing or obstacles.

Another significant contribution of Glock pistols is their polymer frame construction. Before Glock's introduction, handguns were predominantly made from metal frames, which often resulted in heavier and bulkier firearms. However, Gaston Glock recognized the potential of using polymer materials to construct gun frames, resulting in a lighter yet durable alternative. This innovation not only revolutionized firearm manufacturing but also improved user experience by reducing fatigue during extended shooting sessions.

Glock pistols are known for their exceptional reliability in adverse conditions - a feature that sets them apart from many other handguns on the market. The simplified design allows for fewer parts compared to traditional firearms, reducing potential points of failure or malfunctioning. Additionally, extensive testing under extreme conditions ensures that every Glock pistol can withstand harsh environments without compromising its performance.

Firearm safety has always been at the forefront of Gaston Glock's designs. To enhance user safety further, he introduced several features unique to his pistols. The most notable among these is the Safe Action System (SAS). These triple independent automatic safeties system includes three integrated mechanisms: trigger safety, firing pin safety, and drop safety. These combined safety features ensure that the firearm can only be discharged when the trigger is intentionally pulled, reducing the risk of accidental discharge.

Furthermore, Glock pistols incorporate an innovative loaded chamber indicator and a striker status indicator. These visual cues allow users to easily determine if a round is present in the chamber or if the striker is cocked, providing an additional layer of safety during handling and storage.

Gaston Glock's dedication to continuous improvement and innovation has cemented his legacy in the firearms industry. His pistols have not only become a symbol of reliability but have also redefined what it means to own a safe and user-friendly firearm. By challenging traditional design norms, Gaston Glock broke barriers and set new standards for firearm safety, reliability, and performance with his iconic Glock pistols.

***

When it comes to firearms, few names are as recognizable and influential as Glock. Gaston Glock, an Austrian engineer, revolutionized the world of handguns with his invention of the

Glock pistol. Since its introduction in the early 1980s, the Glock handgun has made a significant impact on law enforcement agencies and military organizations worldwide. One of the key factors that contributed to Glock's success in law enforcement was its reliability.

The design of the Glock pistol incorporated several innovative features that improved functionality and reduced the risk of malfunctions. The use of polymer materials for both the frame and grip made it lightweight yet durable, while its simplified construction minimized parts that could potentially break or jam. This high level of reliability quickly gained the trust and confidence of law enforcement officers who rely on their firearms in life-threatening situations.

Another significant aspect that propelled Glock into law enforcement favor was its versatility. With a wide range of models available, officers had options to choose from based on their specific needs. Whether it was a compact model for concealed carry or a full-size duty weapon with increased capacity, there was a Glock pistol suitable for every situation. This adaptability allowed agencies to standardize their firearms across departments, simplifying training and logistics.

The introduction of polymer-framed handguns also played a crucial role in increasing adoption among law enforcement agencies. Before Gaston Glock's innovation, most pistols featured metal frames that were heavier and less comfortable to carry for extended periods. The use of polymer not only reduced weight but also enhanced ergonomics by providing better grip texture and customizable backstraps to accommodate different hand sizes.

These improvements significantly contributed to officer comfort during long shifts. Furthermore, Glock's commitment to continuous improvement ensured that their handguns evolved with changing needs over time. By actively seeking feedback from end-users – including law enforcement officers – Glock was able to make iterative refinements to their designs. This ongoing dialogue helped address specific requirements and preferences, such as improved trigger pull, ambidextrous controls, and enhanced accessory compatibility.

By consistently delivering updated models that met the demands of the field, Glock cemented its position as a trusted partner for law enforcement agencies. The impact of Glock's innovation on military adoption of handguns has been equally profound. The reliability and versatility that made Glock firearms popular among law enforcement agencies also appealed to military units worldwide. The lightweight design and simplified construction made it easier for soldiers to carry additional ammunition without compromising maneuverability or combat effectiveness.

Furthermore, the modular nature of Glock pistols allowed soldiers to adapt their firearms for various mission requirements easily. Accessories such as optics, suppressors, and lights could be seamlessly integrated into the pistol's frame or slide without requiring extensive modifications or custom gunsmithing. This modularity made it possible for troops to tailor their firearms to specific roles while maintaining a standardized platform across units.

Gaston Glock's invention has had a significant impact on law enforcement and military adoption of handguns.

***

In the world of firearms, one name stands out above all others – Gaston Glock. With his revolutionary designs and unwavering commitment to quality, Glock has become synonymous with excellence in handgun manufacturing. But what is the secret behind his success? What drives him to constantly push the boundaries of innovation and set new standards for the industry? To answer these questions, we must delve into the mind of this enigmatic figure and explore his vision for manufacturing excellence.

At the heart of Gaston Glock's philosophy is a relentless pursuit of perfection. From the very beginning, he recognized that every aspect of manufacturing – from design to production – must be meticulously planned and executed with precision. His attention to detail is legendary, as he personally oversees each step in the process to ensure that every firearm leaving his factory meets his exacting standards.

One key element in Glock's vision for manufacturing excellence is automation. He understood early on that embracing technological advancements would be crucial in maintaining consistent quality while increasing production efficiency. By utilizing state-of-the-art machinery and robotics, Glock was able to streamline their manufacturing processes and reduce human error. This not only resulted in higher productivity but also ensured that each handgun was crafted with utmost precision.

Furthermore, Glock's commitment to continuous improvement has been a driving force behind his success. He firmly believes that there is always room for enhancement, no matter how advanced a product may seem at present. To achieve this constant evolution, he encourages feedback from customers and carefully studies market trends to identify areas where innovation can make a difference. This dedication to staying ahead of the curve has allowed Glock firearms to remain at the forefront of industry standards.

Another crucial aspect of Gaston Glock's vision lies in fostering a culture of excellence within his organization. He firmly believes that an engaged workforce is instrumental in achieving manufacturing greatness. To this end, he invests heavily in employee training and development, ensuring that each member of his team is equipped with the knowledge and skills necessary to excel in their respective roles.

By empowering his employees, Glock creates an environment where innovation and creativity can flourish.

Moreover, sustainability is a key pillar of Gaston Glock's manufacturing vision. He recognizes the importance of responsible resource management and strives to minimize the environmental impact of his operations. From reducing waste and energy consumption to implementing eco-friendly manufacturing practices, he aims to leave a lasting positive legacy for future generations.

Gaston Glock's vision for manufacturing excellence encompasses a relentless pursuit of perfection, embracing automation and technological advancements, fostering continuous improvement, nurturing a culture of excellence within his organization, and promoting sustainability. Through these principles, he has revolutionized the handgun industry and set new benchmarks for quality and innovation. Gaston Glock's unwavering commitment to these ideals has cemented his position as one of the most influential figures in firearms manufacturing history.

***

When it comes to handguns, one name stands above the rest – Glock. The Austrian firearms manufacturer has achieved unparalleled success and dominance in the global firearms market. With its innovative designs, exceptional reliability, and widespread adoption by law enforcement agencies worldwide, Glock has become the go-to choose for professionals and enthusiasts alike.

At the heart of Glock's success is their commitment to innovation. Gaston Glock, the man behind the brand, revolutionized handgun design with his polymer-framed pistols. Prior to Glock's introduction of these lightweight and durable handguns in the 1980s, most pistols were made primarily from steel. This breakthrough design not only reduced weight but also increased corrosion resistance and improved overall performance.

Another key factor contributing to Glock's dominance is their exceptional reliability. The company prides itself on producing firearms that can function flawlessly under even adverse conditions. Through rigorous testing procedures and quality control measures, each handgun undergoes extensive evaluation before leaving the factory floor. This commitment to excellence ensures that every customer can rely on their Glock pistol when it matters most.

Glock's dedication to simplicity also plays a significant role in their popularity. Unlike many other handguns on the market, Glocks has a straightforward design with fewer components compared to traditional pistols. This simplicity not only makes it easier to disassemble for maintenance but also contributes to their renowned reliability. With fewer parts that can potentially fail or malfunction, Glocks are known for being dependable even in high-stress situations.

Furthermore, Glock handguns have gained widespread adoption by law enforcement agencies worldwide – a testament to their quality and performance. Law enforcement officers demand firearms that they can trust with their life's day in and day out. Time and again, Glocks have proven themselves as reliable tools for these professionals who put themselves at risk protecting communities around the globe.

Glock's rise in popularity among civilian shooters should not be overlooked either. The company's commitment to innovation, reliability, and simplicity has resonated with gun enthusiasts everywhere. Glock pistols are known for their ease of use, comfortable ergonomics, and consistent performance. Whether it's for self-defense, competition shooting, or recreational use, Glock handguns have become the firearm of choice for many.

Lastly, Glock's dominance in the firearms market can be attributed to their commitment to customer service. The company provides extensive support to customers by offering a wide range of accessories and aftermarket parts. Additionally, their customer service representatives are knowledgeable and responsive to inquiries or concerns. This dedication to serving their customers has created a loyal following that further contributes to Glock's global success.

Glock handguns have achieved a global phenomenon status due to their innovative design, exceptional reliability, widespread law enforcement adoption, popularity among civilian shooters, and commitment to customer service. Gaston Glock's vision for producing reliable firearms with cutting-edge technology has revolutionized the firearms industry. As long as there is a demand for quality handguns that can be trusted in any situation – whether by professionals or enthusiasts – it is likely that Glock will continue its dominance in the firearms market for years to come.

Gaston Glock, the visionary behind the world's most popular handgun, has left an indelible mark on the firearms industry. His innovative designs and unwavering commitment to quality have forever changed the landscape of gun manufacturing. As we delve into his remarkable journey, it becomes evident that his influence extends far beyond his own company. Glock's legacy is not only about the success of his brand but also about how he has inspired future generations of gunmakers.

One of Gaston Glock's most significant contributions to the firearms industry lies in his revolutionary approach to handgun design. His creation, the Glock pistol, introduced a groundbreaking polymer frame that revolutionized firearm construction. This design innovation not only reduced weight but also enhanced durability and reliability. The success of this concept led other gunmakers to explore similar avenues in their own designs.

Furthermore, Glock's emphasis on simplicity and functionality set a new standard for handguns. His focus on ergonomics and ease of use resonated with shooters worldwide, making his pistols highly sought-after. This emphasis on user-friendly designs has influenced numerous gunmakers who now prioritize user experience as a key consideration in their product development process.

Another aspect that sets Gaston Glock apart is his commitment to rigorous quality control standards. From inception to production, every Glock firearm undergoes meticulous testing to ensure unparalleled reliability and performance. This unwavering dedication to quality has raised the bar for gun manufacturers across the globe. Gunmakers now understand that meeting or exceeding these exacting standards is crucial in maintaining customer trust and loyalty.

Glock's relentless pursuit of perfection has also inspired innovation in manufacturing techniques within the firearms industry. His ability to leverage cutting-edge technologies and materials pushed boundaries previously thought impossible by many traditionalists in the field. As a result, other gunmakers have followed suit by embracing advanced manufacturing methods such as CNC machining and utilizing modern materials like polymers and composites.

Furthermore, Glock's commitment to continuous improvement has fostered a culture of innovation within the firearms industry. His willingness to adapt and evolve his designs in response to customer feedback and changing market demands has set a precedent for other gunmakers. This emphasis on innovation has led to the development of new features and improvements in firearm technology, benefiting both professional users and civilian shooters alike.

Beyond his technical contributions, Gaston Glock's business acumen and entrepreneurial spirit have also left an enduring impact. His success story serves as an inspiration for aspiring gunmakers, demonstrating that with determination, ingenuity, and perseverance, one can achieve greatness in this fiercely competitive industry.

Gaston Glock's influence on future gunmakers is undeniable. From his revolutionary designs to his unwavering commitment to quality control and innovation, he has set new benchmarks that have inspired countless others in the firearms industry. His legacy will continue to shape the future of gun manufacturing as his innovations are embraced by new generations of firearm enthusiasts worldwide.

***

Throughout history, certain individuals have left an indelible mark on their respective industries, forever changing the landscape. In the realm of firearms, Gaston Glock stands as one such figure, his innovative spirit and relentless pursuit of excellence revolutionizing the world of handguns. From humble beginnings as a manufacturer of curtain rods to creating the iconic Glock pistol, his enduring legacy continues to shape the industry and inspire generations of firearm enthusiasts.

Gaston Glock's impact on the handgun market cannot be overstated. His visionary approach challenged conventional wisdom and brought forth a new era in firearm design. The introduction of polymer-framed handguns marked a significant departure from traditional metal-based pistols, offering enhanced durability, reduced weight, and improved ergonomics. This groundbreaking innovation not only transformed the industry but also set a new standard for reliability and performance.

The Glock pistol quickly gained popularity due to its simplicity and ease of use. Its innovative features like a striker-fired mechanism and lack of external safety switches appealed to both law enforcement agencies and civilian users alike. The gun's modular design allowed for easy customization with interchangeable parts, catering to different shooting preferences and requirements. These qualities propelled Glock pistols to become some of the most sought-after firearms worldwide.

Beyond technological advancements, Gaston Glock's commitment to quality played a crucial role in cementing his legacy. By implementing stringent manufacturing standards and quality control processes within his company, he ensured that every handgun bearing his name would meet or exceed expectations. This dedication to excellence established trust among users who relied on their Glocks for self-defense or professional duties.

Glock's influence extends beyond product development; it encompasses cultural significance as well. The ubiquity of his handguns in popular culture further solidifies their place in history. Countless movies, television shows, and video games have featured characters wielding Glocks – an emblematic representation of power, reliability, and efficiency. This cultural impact not only reinforces the brand's popularity but also serves as a testament to Gaston Glock's lasting influence on modern society.

Furthermore, Gaston Glock's philanthropic endeavors exemplify his commitment to making a positive difference in the world. Through the Glock Foundation, he has supported numerous charitable organizations, focusing on education, medical research, and social welfare. By investing in these initiatives, he demonstrates that his legacy extends beyond firearms – it is about empowering communities and fostering progress.

Gaston Glock's enduring legacy in the world of handguns is undeniable. His innovative designs revolutionized firearm manufacturing and set new industry standards. The widespread adoption of his pistols by both professionals and civilians alike attests to their unmatched performance and reliability. Moreover, through his philanthropic efforts, Glock has shown that he values not only innovation but also making a positive impact on society.

As future generations continue to benefit from his contributions and advancements in handgun technology, Gaston Glock will forever be remembered as a pioneer who reshaped an industry and left an indelible mark on firearms history.

Epilogue

There have been so many firearms pioneers over the years that it was hard to narrow it down to the ones I covered. Multiple chapters ended up on the cutting room floor. I tried to give an interesting cross section of designers, but there are more out there waiting to be discovered.

The next John Browning will not be from a traditional firearms background. They will be a 3D designer using a CAD program to turn out a truly revolutionary design using a 3D printer. The future is bright, and you cannot stop the signal.

Made in the USA
Middletown, DE
14 October 2023

40541562R00292